The Material Basis of Energy Transitions

The Material Basis of Energy Transitions

Edited by

Alena Bleicher

Alexandra Pehlken

ACADEMIC PRESS

An imprint of Elsevier

Academic Press is an imprint of Elsevier
125 London Wall, London EC2Y 5AS, United Kingdom
525 B Street, Suite 1650, San Diego, CA 92101, United States
50 Hampshire Street, 5th Floor, Cambridge, MA 02139, United States
The Boulevard, Langford Lane, Kidlington, Oxford OX5 1GB, United Kingdom

Library of Congress Cataloging-in-Publication Data
A catalog record for this book is available from the Library of Congress

British Library Cataloguing-in-Publication Data
A catalogue record for this book is available from the British Library

ISBN: 978-0-12-819534-5

For information on all Academic Press publications
visit our website at https://www.elsevier.com/books-and-journals

Publisher: Brian Romer
Acquisitions Editor: Graham Nisbet
Editorial Project Manager: Sara Valentino
Production Project Manager: Omer Mukthar
Cover Designer: Mark Rogers

Typeset by SPi Global, India

Contents

Contributors xi

CHAPTER 1 **The material basis of energy transitions—An introduction** 1
Alena Bleicher and Alexandra Pehlken

CHAPTER 2 **The repatriation challenge: Critical minerals mining in the United States** 11
Roopali Phadke

Introduction 11
The US context 11
Reviewing the literature on extraction and society 14
Methods 16
Stories from the mining fields 17
    Alaska 17
    Wyoming 18
    Texas 19
    Critical elements and social license 21
An emerging social science research agenda 22
References 23

CHAPTER 3 **Metal-energy nexus in the global energy transition calls for cooperative actions** 27
Peng Wang, Nan Li, JiaShuo Li, and Wei-Qiang Chen

Introduction 27
Global energy system shifts from carbon driven to metal driven 28
    Energy transition: Becoming more metal driven? 28
    Energy-critical elements: Nexus energy and critical metals 31
The sustainable material cycle in the global energy transition 32
Global challenges call for cooperative action 40
    Challenges: Pressing constraints beyond physical scarcity 40
    Looking ahead: Joint efforts by stakeholders throughout the metal cycle and trade network 43
Acknowledgments 44
References 45

CHAPTER 4   The dependency of renewable energy technologies on critical resources 49

Volker Zepf

Introduction to the "critical," "resources," and "renewables" 49
   Renewable energy technologies..............................50
   The "criticality" aspect 51
   Material and nonmaterial resources.....................53
Renewable energies and their material resources demand 54
   Biomass...55
   Water...56
   Solar..59
   Wind energy..62
   Geothermal energy..65
Conclusion: Materials demand for renewable energies.............66
References...68

CHAPTER 5   Stationary battery systems: Future challenges regarding resources, recycling, and sustainability 71

Marcel Weil, Jens Peters, and Manuel Baumann

Background...71
Future demand for stationary energy storage 72
Potential lithium-based electrochemical energy storage options..72
Calculation of the potential resource demand for batteries 74
   Approach, assumptions, and scenarios 74
   Discussion of the scenarios.......................................79
Recycling and sustainability aspects of LIBs and PLBs..............81
   Recycling of LIBs..81
   PLBs and potential recycling challenges 82
Conclusions..85
Acknowledgment 86
References...86

CHAPTER 6   Making critical materials valuable: Decarbonization, investment, and "political risk" 91

Paul Robert Gilbert

Introduction..91
   "Resource-making" and critical raw materials 92

Valuation, critical raw materials, and "political risk"..................95
"Strategic" minerals, royalties, and stabilization provisions
in DRC..97
HySA and South Africa's platinum futures 101
Conclusion: Provincializing criticality assessments...................104
References...105

**CHAPTER 7   Environmental impacts of mineral sourcing
and their impacts on criticality** **109**
Benjamin C. McLellan

Introduction...109
Environmental impacts—Scope and data....................................110
   The lifecycle perspective 110
   Local environmental impact perspective 111
   Energy and environmental impacts.......................................113
Changing environmental impacts and mitigation 114
   Mines 114
   Processing, extraction, smelting, and refining 116
   Demand side 117
   Waste and recycling 117
Conclusions...118
References...118

**CHAPTER 8   Limits of life cycle assessment in the context
of the energy transition and its material basis** **121**
Fernando Penaherrera and Alexandra Pehlken

Introduction...121
Life cycle assessment (LCA) and resource consumption
indicators..123
   The role of indicator selection in LCA 126
   Resource accounting methods (RAM)....................................127
   Midpoint methods..128
   Endpoint methods 129
   What is the right indicator or method? 130
Indicators and their relationship to the material demands
of emerging renewable energy technologies.............................132
Discussion on the use of indicators for evaluating resource
saving potentials 133
Conclusion 134
References...136

CHAPTER 9   Critical resources, sustainability, and future generations **141**

Björn Koch

Introduction 141

Critical resources, conflict resources, and sustainability 142

Critical resources, conflict resources, and future generations 145

Parfit's paradox and future generations 146

The needs of future generations 148

How many future generations do we need to consider? 149

Taking precautions for future generations 149

Conclusions 150

References 151

CHAPTER 10   Conflicts related to resources: The case of cobalt mining in the Democratic Republic of Congo **153**

Louisa Prause

Introduction 153

Analyzing mining conflicts from a political-ecological
perspective 155

Cobalt: From the DRC to the electric car 156

Conflicts around cobalt mining: The upstream end
in the DRC 159

Conflicts around cobalt mining: The downstream end
in the Global North 160

Conflicts around cobalt mining: Connections and
contradictions 162

Conclusion: Conflicts linked to raw materials required
for energy transitions 163

References 164

CHAPTER 11   Voluntary sustainability initiatives: An approach to make mining more responsible? **169**

Gudrun Franken, Laura Turley, and Karoline Kickler

Introduction 169

The landscape of voluntary sustainability initiatives 172

Overview of initiatives 172

Content of sustainability requirements 176

Assurance systems 178

Achievements and limitations 178

Reach of voluntary initiatives 178

Harmonization of standard requirements 179

Transparency of mining and mineral supply chains 180
Sustainability reporting and performance 182
Limited capacity to deal with deviant and illegal
behavior 182
The cost of certification and voluntary initiatives 183
Conclusion 183
Acknowledgments 184
References 184

**CHAPTER 12 The role of a circular economy for energy
transition** **187**
James R.J. Goddin

Introduction 187
Resource supply and competition 190
Reducing supply risks using circular economy 191
Product design and materials recovery 194
Conclusions 196
References 197

**CHAPTER 13 Substitution of critical materials, a strategy to deal
with the material needs of the energy transition?** **199**
James R.J. Goddin

Material substitution for renewable energy technologies 199
Substance for substance substitution 199
Service for product substitution 202
Process for process substitution 203
New technology for substance 203
Creating substitutes: By engineering simulation 204
Barriers to substitution 205
Conclusions 205
References 206

**CHAPTER 14 Renewable energy technologies and their
implications for critical materials from a sociology
of consumption perspective: The case of
photovoltaic systems and electric vehicles** **207**
Marco Sonnberger

Introduction 207
Consumption and society-nature relations 208
The "why" of consumption: Sociocultural and
sociopsychological functions of goods and services 212
The case of electric vehicles and photovoltaic systems 214

Discussion and conclusion...216
References..218

**CHAPTER 15 Renewable energy and critical minerals: A field
worthy of interdisciplinary research**...........................**223**
Alexandra Pehlken and Alena Bleicher

Author Index 229
Subject Index 239

Contributors

Manuel Baumann
Institute for Technology Assessment and Systems Analysis (ITAS)/Karlsruhe Institute of Technology, Karlsruhe, Germany

Alena Bleicher
Helmholtz Centre for Environmental Research—UFZ, Leipzig, Germany

Wei-Qiang Chen
Institute of Urban Environment, Chinese Academy of Sciences, Xiamen, China

Gudrun Franken
Bundesanstalt für Geowissenschaften und Rohstoffe, Hannover, Germany

Paul Robert Gilbert
International Development, University of Sussex, Brighton, United Kingdom

James R.J. Goddin
Granta Design, Rustat House, Cambridge, United Kingdom

Karoline Kickler
Bundesanstalt für Geowissenschaften und Rohstoffe, Hannover, Germany

Björn Koch
COAST—Centre for Environment and Sustainability, Carl von Ossietzky University of Oldenburg, Oldenburg, Germany

JiaShuo Li
Institute of Blue and Green Development, Shandong University, Weihai, China

Nan Li
Institute of Urban Environment, Chinese Academy of Sciences, Xiamen, China

Benjamin C. McLellan
Graduate School of Energy Science, Kyoto University, Kyoto, Japan

Alexandra Pehlken
OFFIS—Institute for Information Technology, Oldenburg, Germany

Fernando Penaherrera
Carl von Ossietzky University of Oldenburg, Oldenburg, Germany

Jens Peters
Helmholtz Institute Ulm for Electrochemical Energy Storage (HIU), Ulm, Germany

Roopali Phadke
Department of Environmental Studies, Macalester College, St Paul, MN, United States

Louisa Prause
Albrecht Daniel Thaer-Institute of Agricultural and Horticultural Sciences, Agricultural and Food Policy Group, Humbolt University of Berlin, Berlin, Germany

Marco Sonnberger
ZIRIUS—Research Center for Interdisciplinary Risk and Innovation Studies, University of Stuttgart, Stuttgart, Germany

Laura Turley
International Institute for Sustainable Development & University of Geneva, Geneva, Switzerland

Peng Wang
Institute of Urban Environment, Chinese Academy of Sciences, Xiamen, China

Marcel Weil
Institute for Technology Assessment and Systems Analysis (ITAS)/Karlsruhe Institute of Technology, Karlsruhe; Helmholtz Institute Ulm for Electrochemical Energy Storage (HIU), Ulm, Germany

Volker Zepf
Independent consultant for Geography and Resource Strategies, Augsburg, Germany

The material basis of energy transitions—An introduction

1

Alena Bleicher[a] and Alexandra Pehlken[b]
[a]Helmholtz Centre for Environmental Research—UFZ, Leipzig, Germany
[b]OFFIS—Institute for Information Technology, Oldenburg, Germany

This book investigates the interdependencies of renewable energy systems and their material basis during the various life-cycle phases of the energy technologies involved, with a focus on scarce and critical raw materials. Specific raw materials such as neodymium, lithium, and cobalt are required to produce renewable energy systems, most notably energy technologies such as wind turbines, photovoltaic cells, and batteries. Raw materials are extracted from geological repositories or become available in the market through recycling in the end-of-life phase of objects and technologies. Thus, renewable energy systems depend on either mining operations or recycling efforts, with the systems themselves becoming "urban mines" when energy infrastructures are decommissioned.

The idea for this book emerged during a conference where we, the editors of this book, Alexandra Pehlken and Alena Bleicher, met as leaders of research groups funded by the FONA research program (Research for Sustainable Development), which was initiated by the German Federal Ministry of Education and Research. Although we have different disciplinary backgrounds—engineering and social science—we share an interest in the provision of raw materials for advanced technological applications. The research group *Cascade Use*, led by Alexandra Pehlken, conducted research from an engineering perspective on issues such as the cascading use of materials, using case studies in the automotive and renewable energy sectors. The cascading use of raw materials was assessed across more than one life cycle, e.g., a lithium ion battery intended for use in cars that is reused for the stationary storage of energy within the grid. At its end-of-life phase, the battery will be recycled and its raw materials will enter the raw material market. The research group *GORmin*, led by Alena Bleicher, aimed to explain how the development of new technologies for exploiting, extracting, and processing resources from geological or anthropogenic repositories is shaped by societal factors, such as the practices and daily decision-making routines in environmental administration and research projects, as well as conflict dynamics, and regional mining histories and narratives. The concept of socio-technical systems underlies this research. This concept views technical and social systems as interrelated—it proposes that technology shapes society and vice versa.

The Material Basis of Energy Transitions. https://doi.org/10.1016/B978-0-12-819534-5.00001-5

During our research, we realized that studies related to the so-called "critical raw materials" are often legitimized by references to renewable energy technologies (see, e.g., Sovacool et al., 2020). However, a closer look revealed that the interrelation between these two fields is often ignored, and has not been systematically or comprehensively considered in scientific research. Within the last decade, research has been carried out by scientists with diverse scientific backgrounds (e.g., geology, engineering, industrial ecology, geography, sociology, anthropology) on issues related to *either* renewable energy systems and technologies *or* mining and the processing of (specific) minerals. Many books and scientific papers have shed light on the methods, challenges, and impacts of energy transitions on societies (e.g., Chen, Xue, Cai, Thomas, & Stückrad, 2019; Cheung, Davies, & Bassen, 2019; Dietzenbacher, Kulionis, & Capurro, 2020; Viebahn et al., 2015). A broad range of issues related to renewable energy systems have been discussed: secure and stable energy provision (e.g., Sinsel, Riemke, & Hoffmann, 2020), political strategies to support renewables (e.g., market incentives, regulations) (e.g., Overland, 2019; Verbong & Loorbach, 2012), the impact of energy transformation on social justice (e.g., Simcock, Thomson, Petrova, & Bouzarovski, 2017), the energy-food nexus in the context of bioenergy (e.g., Levidow, 2013; Wu et al., 2018), perceptions of and conflicts related to renewable energy technologies (e.g., Benighaus & Bleicher, 2019; Rule, 2014; Truelove, 2012), and the challenges of managing (smart) grids (e.g., Hossain et al., 2016; Smale, van Vliet, & Spaargaren, 2017).

The issue of nonenergetic raw material provision is almost exclusively debated in the fields of raw materials—resource policy, industry, and science. Recently, working papers and journal articles have discussed the issue of secure supply chains for specific raw materials that are used to produce advanced technology (e.g., Blagoeva, Aves Dias, Marmier, & Pavel, 2016; Langkau & Tercero Espinoza, 2018; Løvik, Hagelüken, & Wäger, 2018), as well as problems related to mining such as conflicts over resources (e.g., Kojola, 2018; Martinez-Alier, 2009), environmental, health and security issues in small-scale artisanal mining (e.g., Jacka, 2018; Smith, 2019), and the potential and limits of management instruments in mining (e.g., Owen & Kemp, 2013; Phadke, 2018).

In order to address the challenges related to the energy transition and its material basis, a broader perspective must be taken. First, it is necessary to consider the interdependencies of renewable energy systems, their future development, technology paths, resource extraction, and resource provision. Second, these relationships have to be explored from different disciplinary angles in order to identify potentially problematic aspects. Thus, questions of global justice, responsible mining and consumption, and the effects of price volatility need to be considered together with energy and climate policies, scenarios for future development, technological questions about innovative technologies in different fields of energy use and provision (electricity, heat, traffic), alternative resources (e.g., recycling potentials), as well as investment strategies developed by industry and policymakers to address the challenges.

This book aims to provide a comprehensive interdisciplinary overview of issues related to decentralized renewable energy systems and their mineral basis, and to gather together previously unrelated perspectives from natural sciences, engineering, and social sciences. By doing so, the book serves those who are interested in a raw material demand perspective on the energy transition and renewable energy. Our readers will likely be scientists from diverse disciplines and professionals in different fields of work, such as business and industry, finance, and public policy. The book is suitable for people with no prior knowledge of these issues, such as undergraduate and graduate students, as well as experts in related fields, who will find valuable reflections and inspiration for future research.

In this book, we have assembled contributions from authors who have already researched the relationship between renewable energy technologies, energy systems, and the material basis, or who have research experience in one of these areas and were willing to take on a dual perspective for this book. The authors discuss a range of issues. We have briefly summarized them here to give readers some guidance about the structure and content of the book.

Several authors aim to more precisely characterize the scale of the problem and the dynamics of the issue by describing the type and amount of minerals needed for energy systems. By taking a historical perspective, Peng Wang and his colleagues (Chapter 3) show how the global energy system's demand for and consumption of materials has increased and diversified within the last few decades. Wang et al., Zepf (Chapter 4), and Goddin (Chapter 13) explain that one reason for this diversification is that the materials in question provide specific technological services. For instance, elements such as gallium, germanium, and indium are used in thin film photovoltaics, as they have a high absorption coefficient and are extremely effective at absorbing sunlight. Rare-earth elements such as neodymium and rhenium are used in permanent magnets, as they have a high curie temperature (the temperature at which magnetization is lost), and are resistant to corrosion.

Wang et al. (Chapter 3), Zepf (Chapter 4), and Weil et al. (Chapter 5) all start by specifying the materials required for energy technologies. These include generation technologies such as wind and solar systems, as well as storage technologies such as batteries. Using different approaches (e.g., material flow analyses, scenario analyses), these authors then determine the amount of materials needed for an energy system based on renewable energy. Peng Wang et al. present a mineral-energy nexus framework to assess material demand, and the flow and stocks along the material cycle. They categorize energy technologies into wind- or motor-related technology, photovoltaic-related technology, battery technology, and vehicle-related technology, and use these categories as entry points for their discussion of the challenges posed by the system of international trade, and environmental issues related to the provision of relevant materials. Based on the state of the art of renewable energy technologies and expectations regarding their future development, Volker Zepf provides an overview of the amount of resources needed for the production of energy from biomass, hydro, solar, wind, and geothermal resources. He concludes that some wind and solar technologies will require high amounts of "critical materials." In addition, Marcel Weil and his colleagues discuss the resources required for stationary battery systems. They consider the material and environmental consequences of a scenario in which the global transition to an electricity system based on 100% renewable energy is achieved by 2050. The authors of these chapters point out the importance of differentiating between the notions of "resources" and "reserves" when estimating the availability of a given mineral. A resource is a concentration of minerals that has likely prospects of economic recovery in the future. Reserves are concentrations of minerals that can be recovered and processed today in a technically and economically feasible way, and which are legally accessible, meaning that someone has legal permission to extract the minerals (BGS, British Geological Survey, 2019).

A central concept regarding the material basis of renewable energy systems is "criticality" or "critical materials." While the abovementioned authors rely on notions of criticality used by bodies such as the European Union, others critically discuss the concept and its current use, and highlight its shortcomings. From a science and technology studies perspective, Paul Gilbert (Chapter 6) reveals assumptions, resource imaginaries, and measures that are embedded in the concept of criticality, and which are built upon future energy scenarios. Based on his findings, he provides a fundamental critique of these entanglements. Gilbert shows that instruments such as political risk assessments are based on the needs of wealthy resource-importing countries, and that these instruments risk reproducing colonial relationships, as well as neglecting local and national aspects that are relevant in mining countries. Indeed, Wang et al. (Chapter 3), Phadke (Chapter 2), and Gilbert (Chapter 6) all demonstrate the influence that national political decisions (e.g., resource or energy policies) have on geopolitical power constellations and whether or not minerals are viewed as "critical."

Other authors criticize the limited scope and economic focus of criticality definitions and assessments: Wang et al. (Chapter 3), McLellan (Chapter 7), and Koch (Chapter 9). These authors argue that environmental impacts along the product chain must also be taken into consideration during criticality assessments. Björn Koch reveals that such assessments currently neglect both ecological and social aspects. In his chapter, he takes a closer look at the concepts of "critical resources" and "conflict resources," and relates them to notions of sustainability and sustainable development. Based on these notions, he identifies the moral obligations intertwined with the handling and consumption of resources. Koch also clarifies the differences between "critical materials" and "conflict minerals": the former relies exclusively on economic considerations, while the latter is derived from human rights and international law, and focuses on moral obligations toward all human beings.

Several authors show that life-cycle assessment (LCA) approaches could potentially be used to assess, evaluate, describe, and quantify the criticality of resources in order to provide knowledge for (political) decision-making. McLellan (Chapter 7), Penaherrera and Pehlken (Chapter 8), and Weil et al. (Chapter 5) discuss the limits and shortcomings of LCA approaches currently in use, and suggest possible improvements. From an environmental impact assessment perspective, Benjamin McLellan emphasizes the relevance of local environmental aspects, most notably the impacts of mining on water usage, land usage, and pollution. He criticizes the fact that these aspects are not considered in criticality assessments, even though there are methods available that would facilitate the incorporation of environmental factors (e.g., environmental impact assessment). Fernando Penaherrera and Alexandra Pehlken point out a major problem with LCA approaches: The results of LCAs often cannot be compared, because the assessments use different indicators and specify different system borders. Based on an overview of existing methods in LCA and the indicators used for the assessment of raw material consumption, the authors discuss further limitations of these approaches that occur when they are used to evaluate the impacts of implementing new energy technologies. Their analysis reveals that LCA methods currently focus on aspects such as material depletion or global warming potential. Instead, the authors argue, LCA should investigate aspects such as material criticality (the scarcity of certain minerals), material efficiency, and the potential for replacement. Marcel Weil et al. suggest that LCA should be expanded to include the potential of recycling technologies for resource provision.

Mining and the extraction of resources from geological repositories will also be central for the provision of raw materials in the future. Several authors highlight current trends in mining and their effects. Benjamin McLellan (Chapter 7) makes us aware that the shift from coal to renewables is leading to a shift in mining locations due to the geological availability of specific resources. Furthermore, new mineral requirements are leading to more complex and more energy-intensive mining processes, which are designed to handle lower grade ores, and repositories located in greater depths. These trends are causing dramatic changes in land use patterns and pose risks for the environment (e.g., water use and pollution). Other authors discuss the effects of market dynamics, political strategies, and technological developments on mining and the formation or dissolution of monopolies. Goddin and Zepf reveal that long-term expertise and technology for extracting and refining rare-earth metals are some of the reasons behind China's monopoly in this field. Wang et al. argue that monopolies are shifting due to new trends in technology, such as the rise of electro-mobility. Political strategies such as diversification and risk spreading in resource supply (Goddin, Chapter 12), trade protectionism (Wang et al.), allocation of criticality status (Gilbert), and repatriation (Phadke) are impacting mining activities and have led to the opening and closure of mines.

Another trend related to and potentially impacting mining activities is the emergence of regulations and specific legislation (Goddin, Chapter 12). Gudrun Franken and her colleagues (Chapter 11) provide an overview of voluntary initiatives, standards, and certificates that have been developed in order to evaluate the sustainability of practices in mining and the supply chains of base metals, industrial minerals, and rare-earth elements. Franken et al. analyze 19 of these initiatives and discuss their differences with regard to the sustainability aspects they cover. The authors conclude that the existing standards do not sufficiently address minerals that are important for renewable energies. Other shortcomings include gaps between sustainability reporting and performance, and the fact that certain mining regions tend to produce un-certified minerals. The authors argue that the observable trend of harmonization and cross-acknowledgment of standards should be welcomed, because this may streamline procedures in a way that might finally lead to a single internationally recognized framework of reference standards. Franken et al. interpret the emergence of these initiatives as evidence of an awareness that mining has to become more responsible. Roopali Phadke (Chapter 2) notes that the concept of responsible mining is currently monopolized by corporations and their political champions, and doubts that it serves the most vulnerable parties unless policymakers and citizens' groups in mining regions are involved in defining what "responsible mining" actually means.

This hints at friction between global, national, and regional actors, which is highly relevant in the context of renewable energy technologies and their material basis. Several authors refer to this friction in their chapters. Louisa Prause (Chapter 10) looks at the example of cobalt mining in the Democratic Republic of Congo (DRC). She reveals that conflicts related to the mining of metals for green technologies occur not only upstream in the product chain between artisanal miners and industrial mining in the DRC, but also downstream in Northern countries, where initiatives criticize industries for disregarding the unsafe and unjust cobalt production conditions. She demonstrates that the number of conflicts downstream and upstream increased and diversified due to the rise in demand for cobalt linked to e-mobility strategies. Prause emphasizes the opportunities and challenges of combining local initiatives in resource-producing countries, such as trade unions in the DRC, with initiatives in resource-consuming countries. By tracing current trends in global and US policy, Roopali Phadke (Chapter 2) shows the tensions caused by the entanglement of clean energy policies and mining advocacy in the United States. Using three case studies from the United States, she reveals how national and federal mining policies that disregard local interests (often represented by citizens' groups) have led to conflicts. She concludes that long-term climate adaptation priorities have to accommodate the environmental justice concerns of local communities. In similar vein, Benjamin McLellan highlights the importance of understanding the local factors affecting current and future mines, such as environmental impacts and land use, for the provision of minerals. Paul Gilbert (Chapter 6) also discusses the tension between perspectives from Northern countries and resource-rich, postcolonial countries, arguing that the political and social perspectives of the latter need to be taken into consideration.

Almost all the contributing authors refer to recycling and recovery as promising approaches for the provision of secondary resources for renewable energy technologies (Phadke, Penaherrera and Pehlken, Koch, Sonnberger, Weil et al., Wang et al.). However, the authors also point to numerous problems related to the recycling of critical materials, such as limited economic feasibility (Zepf, Wang et al., Sonnberger), a lack of recycling infrastructure and technologies (Zepf, McLellan, Prause), limited availability of material stocks (Zepf), and high energy demands (Penaherrera and Pehlken). In response, James Goddin (Chapter 12) discusses the potential and challenges of "circular economies": a concept that may help to solve some of these issues. He highlights the main principles of the circular economy approach: decoupling growth from consumption, making products more durable and extending their useful service life, as well as new forms of consumption such as product sharing (e.g., cars). He argues that this concept is not only beneficial for the environment and suitable for dealing with material supply risks, but also fits the corporate governance structures of mainstream companies, because it links to the revenue and risk metrics that are deeply rooted in most corporate cultures. However, circular economies also require a change in the mindset of engineers and designers, who have to consider the recyclability of renewable energy technologies (Goddin (Chapter 12), Penaherrera and Pehlken, Weil et al.). Goddin makes it clear, as do several other authors in their chapters, that a circular economy (and recycling) is not a silver bullet, but can be part of a solution for the current challenges in raw material provision.

Against this background, several authors highlight the potential of another promising approach: the substitution of scarce materials (Penaherrera and Pehlken, Koch). James Goddin (Chapter 13) describes four types of substitution: substance for substance, service for product, process for process, and new technology for substance. He discusses the possibility of substituting materials with others, concluding that the development of substitutes is challenging and takes a long time, although this process can be supported by simulation tools. Substitutes often involve compromising on performance, cost or reliability, and they usually have to be tested using extensive procedures before they can be deemed technically and legally reliable.

A constant theme throughout the book is the role that consumers play, or can play, in the relationship between energy technologies and their constituent materials. While some authors identify end consumers and the difficulty of changing dominant consumption patterns as stumbling blocks in the debate (e.g., Zepf, Phadke), others see consumers as important actors whose decisions and engagement can initiate major changes. Goddin (Chapter 12) argues that consumers increasingly expect the provision of low-carbon technologies, and are willing to spend more money on socially and ecologically favorable products and services (e.g., sharing products). Prause shows that consumer campaigns initiated by civil society organizations in Western countries encourage consumers to exert pressure on the producers, and to demand the responsible sourcing of minerals, for instance, or that car manufacturers be transparent about their supply chains. Similarly, Franken et al. explain how certification schemes are designed with end consumers in mind, although "labels to inform end consumers about responsible production are still rare for minerals and metals." Against this background, Marco Sonnberger adds the perspective of consumption and consumers (Chapter 14). Sonnberger looks at electric vehicles (EV) and photovoltaic installations (PV) to discuss the role consumers have or might have regarding the use of critical materials. His analysis is based on concepts from consumption studies, and he concludes that consideration of the materials used is most relevant during the production phase, and is consequently beyond the influence of consumers. By referring to existing research, he shows that consumers who invest in PV or EV are seldom ecologically minded, but instead make their decision within a system of energy policies and market incentives.

Together, the contributions in this book shed light on all the phases of the renewable energy technology life cycle, from the resource extraction and the use phase to the end-of-life phase and recycling. By choosing a particular conceptual perspective, and by referring to different case studies, each chapter deals with a specific issue that characterizes the interdependency of critical raw materials and renewable energies.

References

Benighaus, C., & Bleicher, A. (2019). Neither risky technology nor renewable electricity: Contested frames in the development of geothermal energy in Germany. *Energy Research & Social Science, 47*, 46–55. https://doi.org/10.1016/j.erss.2018.08.022.

BGS, British Geological Survey. (2019). *What is the difference between resources and reserves?* https://www.bgs.ac.uk/mineralsuk/mineralsYou/resourcesReserves.html (Accessed December 8, 2019).

Blagoeva, D. T., Aves Dias, P., Marmier, A., & Pavel, C. C. (2016). Assessment of potential bottlenecks along the materials supply chain for the future deployment of low-carbon energy and transport technologies in the EU. In *Wind power, photovoltaic and electric vehicles technologies, time frame: 2015-2030, JRC Science for Policy Report*. https://setis.ec.europa.eu/sites/default/files/reports/materials_supply_bottleneck.pdf (Accessed December 17, 2019).

Chen, C., Xue, B., Cai, G., Thomas, H., & Stückrad, S. (2019). Comparing the energy transitions in Germany and China. Synergies and recommendations. *Energy Reports, 5*, 1249–1260. https://doi.org/10.1016/j.egyr.2019.08.087.

Cheung, G., Davies, P. J., & Bassen, A. (2019). In the transition of energy systems: What lessons can be learnt from the German achievement? *Energy Policy, 132*, 633–646. https://doi.org/10.1016/j.enpol.2019.05.056.

Dietzenbacher, E., Kulionis, V., & Capurro, F. (2020). Measuring the effects of energy transition: A structural decomposition analysis of the change in renewable energy use between 2000 and 2014. *Applied Energy, 258*, 114040. https://doi.org/10.1016/j.apenergy.2019.114040.

Hossain, M. S., Madlool, N. A., Rahim, N. A., Selvaraj, J., Pandey, A. K., & Khan, A. F. (2016). Role of smart grid in renewable energy: An overview. *Renewable and Sustainable Energy Reviews, 60*, 1168–1184. https://doi.org/10.1016/j.rser.2015.09.098.

Jacka, J. K. (2018). The anthropology of mining: The social and environmental impacts of resource extraction in the mineral age. *Annual Review of Anthropology, 47*, 61–77. https://doi.org/10.1146/annurev-anthro-102317-050156.

Kojola, E. (2018). Bringing back the mines and a way of life: Populism and the politics of extraction. *Annals of the American Association of Geographers, 109*(2), 371–381. https://doi.org/10.1080/24694452.2018.1506695.

Langkau, S., & Tercero Espinoza, L. A. (2018). Technological change and metal demand over time: What can we learn from the past? *Sustainable Materials and Technologies, 16*, 54–59. https://doi.org/10.1016/j.susmat.2018.02.001.

Levidow, L. (2013). EU criteria for sustainable biofuels: accounting for carbon, depoliticising plunder. *Geoforum, 44*(1), 211–223. https://doi.org/10.1016/j.geoforum.2012.09.005.

Løvik, A. N., Hagelüken, C., & Wäger, P. (2018). Improving supply security of critical metals: Current developments and research in the EU. *Sustainable Materials and Technologies, 15*, 9–18. https://doi.org/10.1016/j.susmat.2018.01.003.

Martinez-Alier, J. (2009). Social metabolism, ecological distribution conflicts, and languages of valuation. *Capitalism Nature Socialism, 20*(1), 58–87. https://doi.org/10.1080/10455750902727378.

Overland, I. (2019). The geopolitics of renewable energy: Debunking four emerging myths. *Energy Research & Social Science, 49*, 36–40. https://doi.org/10.1016/j.erss.2018.10.018.

Owen, J. R., & Kemp, D. (2013). Social licence and mining: A critical perspective. *Resources Policy, 3*(8), 29–35. https://doi.org/10.1016/j.resourpol.2012.06.016.

Phadke, R. (2018). Green energy futures: Responsible mining on Minnesota's Iron Range. *Energy Research & Social Science*, *35*, 163–173. https://doi.org/10.1016/j.erss.2017.10.036.

Rule, T. A. (2014). *Solar, wind, and land: Conflicts in renewable energy development*. London: Routledge.

Simcock, N., Thomson, H., Petrova, S., & Bouzarovski, S. (Eds.), (2017). *Energy Poverty and Vulnerability. A Global Perspective*. London: Routledge.

Sinsel, S. R., Riemke, R. L., & Hoffmann, V. H. (2020). Challenges and solution technologies for the integration of variable renewable energy sources—a review. *Renewable Energy*, *145*, 2271–2285. https://doi.org/10.1016/j.renene.2019.06.147.

Smale, R., van Vliet, B., & Spaargaren, G. (2017). When social practices meet smart grids: Flexibility, grid management, and domestic consumption in The Netherlands. *Energy Research & Social Science*, *34*, 132–140. https://doi.org/10.1016/j.erss.2017.06.037.

Smith, N. M. (2019). "Our gold is dirty, but we want to improve": Challenges to addressing mercury use in artisanal and small-scale gold mining in Peru. *Journal of Cleaner Production*, *222*(10), 646–654. https://doi.org/10.1016/j.jclepro.2019.03.076.

Sovacool, B. K., Ali, S. H., Bazilian, M., Radley, B., Nemery, B., Okatz, J., et al. (2020). Sustainable minerals and metals for a low-carbon future. *Science*, *6473*(367), 30–33. https://doi.org/10.1126/science.aaz6003.

Truelove, H. B. (2012). Energy source perceptions and policy support: Image associations, emotional evaluations, and cognitive beliefs. *Energy Policy*, *45*, 478–489. https://doi.org/10.1016/j.enpol.2012.02.059.

Verbong, G. & Loorbach, D. (Eds.), (2012). *Governing the energy transition: Reality, illusion or necessity?* London: Routledge.

Viebahn, P., Soukup, O., Samadi, S., Teubler, J., Wiesen, K., & Ritthoff, M. (2015). Assessing the need for critical minerals to shift the German energy system towards a high proportion of renewables. *Renewable and Sustainable Energy Reviews*, *49*, 655–671. https://doi.org/10.1016/j.rser.2015.04.070.

Wu, Y., Zhao, F., Liu, S., Wang, L., Qiu, L., Alexandrov, G., et al. (2018). Bioenergy production and environmental impacts. *Geoscience Letters*, *5*(14), https://doi.org/10.1186/s40562-018-0114-y.

The repatriation challenge: Critical minerals mining in the United States

2

Roopali Phadke
Department of Environmental Studies, Macalester College, St Paul, MN, United States

Introduction

The global conversation about mitigating climate change is now driven by an almost singular focus on deep decarbonization or the "electrification of everything." Many believe that our future lives are powered by stored electricity, which is derived from renewable sources. One of the outstanding challenges of the Anthropocene is balancing our global dream to electrify everything with the brutal fact that clean energy technologies depend on critical minerals mined using nearly medieval techniques. This issue is vexing because critical minerals, like rare earth elements, are vital for four important clean energy applications: permanent magnets for wind turbines and electric cars, advanced batteries, thin-film PV systems, and rare earth phosphors for high-efficiency fluorescent lighting (for a discussion of the criticality concept see Wang et al., Zepf, and Gilbert within this book).

This chapter explores the energy-climate-metals nexus by providing an overview of the challenges for the supply of critical minerals in the United States. I describe how the Trump Administration has advanced mineral extraction and the local political and social challenges that limit the ability of mining companies to extract new materials. I draw on the work of anthropologists, geographers, and science and technology studies scholars to explore how this new era of mining can be understood within a framework of socially constructed metals scarcity. The analysis section describes a range of cases from across the nation where new mines are being permitted and the challenges they have faced. The chapter ends with a broad social science agenda for just energy transitions that place the interest in repatriating mining with other alternatives such as recycling and recovery.

The US context

It is important to view the US critical minerals policy alongside two seemingly contradictory political domains: clean energy policy and mining advocacy. The Obama Administration's promotion of the Paris treaty was a milestone in US

The Material Basis of Energy Transitions. https://doi.org/10.1016/B978-0-12-819534-5.00002-7

climate leadership. The US Paris treaty pledge came in the form of Obama's Clean Power Plan. This was the first comprehensive policy to limit coal energy emissions. The plan aimed to cut the electric sector's carbon pollution by 32% nationally, relative to 2005 levels. US climate policy took an abrupt turn in June 2016, when President Trump gave his now-famous Rose Garden speech where he ceremoniously pulled the United States out of the Paris Accord and cancelled the Clean Power Plan.

After the federal government's withdrawal from climate leadership, the gravity shifted to subnational actors, including states and corporations. The "We are Still In" campaign emerged as a coalition of city mayors, state governors, and corporate leaders to continue to support action to meet the Paris Agreement. Today, across 90 US cities, and more than 10 counties and 2 states, leaders have adopted ambitious 100% clean energy goals (Sierra Club, 2019). In 2018, one of the nation's largest electricity utilities, Xcel Energy, became the first major US utility to pledge to go 100% carbon-free by 2050 and 80% less carbon by 2030.

More recently, American climate policy has been dominated by debates over the Green New Deal. Introduced in the House of Representatives by the fiery new politician, Representative Alexandria Ocasio-Cortez from New York, this ambitious plan builds on President Franklin Roosevelt's New Deal. The new deal was a series of 1930s era programs and projects instituted during the Great Depression to restore prosperity to Americans. The Green New Deal aims to get the United States running on 100% renewable energy by 2030 to avoid the worst consequences of climate change while also trying to fix societal problems like economic inequality and racial injustice. The Green New Deal has been ridiculed by President Trump and his allies. Senator Tom Cotton, a Republican member from Arkansas, said the deal would force Americans to have to "ride around on high-speed light rail, supposedly powered by unicorn tears" (Friedman, 2019).

The New Green Deal, as well as state and corporate pledges to achieve 100% renewable energy, put incredible demand on metals supplies. Trump's trade war with China underscores the need for new domestic supplies, or alternative global suppliers, of critical minerals. Some of these concerns predate Trump's ascension to power. Geopolitical tensions over critical minerals arose in 2010 when China drastically cut 40% of global exports of rare earths in the midst of a territorial dispute with Japan. This shock led many industrialized nations, like Japan and the United States, to look to new supplies and trade alliances, and create domestic critical metals policy (see Wang et al. and Gilbert within this book).

While today the United States imports almost all its rare and precious metals from China, this was not always the case. The United States was the center of rare earths mining for decades. The Molycorp Mountain Pass site, in southern California, operated for nearly 40 years. In 1984, Molycorp met all of the domestic demand for rare earth elements and one-third of the global demand. But as the United States passed stronger environmental laws, the company could not keep up. Mountain Pass was first closed in 2002 due to leaks in its wastewater. The site was reopened for a brief period but eventually closed under bankruptcy in 2015. In 2019, the site was purchased by a Chinese mining consortium, called MP Materials. Today, MP Materials ships its ore to China for processing and has been subject to a 25% tariff since 2019 (Scheyder, 2019).

Even if Mountain Pass had produced at full capacity, the United States needs at least seven mines the size of the Mountain Pass to meet the demand for magnets for wind turbines alone (Clagett, 2013). While the United States has ample domestic supply of critical minerals, particularly in the American West, the main challenge is that a new mine will take an average of 7–10 years to receive state and federal permits, and several additional years for site development before any production begins.

Since Trump's election, both the House and Senate have passed bills to fund more research, fast track projects, and extend loan guarantees to mining companies. The US Congress has focused on passing legislation, like the 2017 METALS Act (H.R. 1407), to provide incentives to industry to develop new sites. Repatriating mining back to the United States has been high on the Trump agenda. In 2017, President Trump signed Executive Order 13817 to boost US domestic supplies of critical metals. Given the patchwork nature of environmental review and mounting opposition, the federal government could play a larger role in providing criteria for siting and permitting. Federal incentives could also be aimed at state agencies to support decision-making and greater public engagement.

The above mining policy uncomfortably intersects with clean energy advocacy. Climate and energy activists have called out the environmental injustices and landscape impacts associated with the nation's dependence on coal mining and natural gas fracking. For example, mountaintop removal for coal mining in Appalachia has resulted in the dramatic cutting of ridges and filling of valleys with disastrous effects on local water quality (Ross, McGlynn, & Bernhardt, 2016). In 2018, US Centers for Disease Control also reported that coal miners in central Appalachia have been disproportionately affected with black lung disease, with as many as 1 in 5 having the affliction. This is the highest level recorded in 25 years (Centers for Disease Control (CDC), 2018).

In these regions, clean energy advocates have also championed wind and solar energy projects. The ironically named Coal River Wind project aimed to construct 220 two-megawatt wind turbines at Coal River Mountain. Yet, as we know now, renewable energy applications have a heavy mining footprint as well. Instead of coal, wind and solar energy would need to exploit other minerals, which while not located in Appalachia will come from someplace and produce human and ecosystem effects. As the United States attempts to rebuild a domestic critical minerals industry, small communities across the nation are caught in the middle of this struggle to address the global challenges of mitigating climate change while dealing with the realities of new mining development.

In a later section of this chapter, I describe case studies of new mines being proposed in the United States in the quest for these critical minerals. In the next section, I first examine how the scholarship from the fields of geography, anthropology, and STS (science and technology studies) help us consider the social production of critical minerals scarcity.

Reviewing the literature on extraction and society

My research on critical minerals development is inspired by a robust social science literature, from anthropologists, geographers, and sociologists, on the impacts of extractive mining on community well-being and ecosystem health. This includes a burgeoning scholarship on mining, energy ethics, and responsible innovation in STS. The following section describes the states of knowledge in these fields and the ways I imagine bringing them into a productive conversation.

Geographers have helped usher in the geologic turn in social theory (Bebbington, 2012; Clark, 2011; Yusoff, 2013). In the much anticipated 2014 publication, *Capitalism and the Earth,* Kathryn Yusoff's manifesto makes a compelling case for reoccupying the strata of the earth. Likewise, in his address at the Association of American Geographers meeting, Tony Bebbington stated that the flurry of new work on minerals and hydrocarbons matters greatly, because they are "constitutive of the functioning of capitalism and when they are enrolled into social life, a wide range of political imaginaries and relationships are reworked" (Bebbington, 2012: 1153). Citing Noel Castree's work, Bebbington further argued that control over the subsoil can be deeply conflictive and signal a sort of Polanyian double movement "where attempts to expand the reach and depth of capitalist commodification are met by vocal (even violent) forms of resistance" (Bebbington, 2012: 1153).

This important new field of "subsoil" social theory is built on decades of work by anthropologists and geographers investigating mining's impacts on communities and the explosion of resistance movements, particularly in the Global South. The more recent applied literature in these fields has investigated how corporate social responsibility programs are emerging in response to both community critique and the need for a public buy-in of local mining operations.

The literature on sustainable mining, corporate social responsibility, and a corporation's "social license to operate" provides insights into the material politics of project development (Benson & Kirsch, 2010; Li, 2011). According to the late ecologist Robert Goodland, whom many considered the World Bank's social conscience, mining is always an issue of social justice, because the impacts on agriculture and the environment are mostly borne by the poor. Goodland argued that "responsible mining" is fundamentally about the design of operations to secure optimal net benefit for the citizens of the host country over the long term with the lowest social and environmental impact. Goodland claimed that obtaining a "social license to mine" requires mining companies to gain the respect, trust, and collaboration of governments and local populations, especially indigenous communities (Goodland, 2012).

While there are no agreed-upon criteria that determine when or how a social license is to be granted, McAllister, Fitzpatrick, and Fonseca (2014) argue that it is a widely accepted principle throughout the mining industry that companies have an obligation to maintain and promote the social and biophysical health of the region affected by their operations. Yet, it is entirely unclear to whom the corporation is responsible; is it to a federal agency, a local regulatory body, citizens groups, or their corporate shareholders? When there is a lack of social resistance to a project, mining corporations assume that this signifies a "social license" to operate rather than a lack of routes for social response (Owen & Kemp, 2013). As Owen and Kemp argue, minimal community resistance is a poor measure of social license because it conflates available evidence of support with "actual" levels of support. While the notion of a social license has been used by industry to suggest its goodwill about the social dimensions of a project, it has rather operated as a way of limiting business risk.

In contrast, Smith Rolston (2015) argued that communities are driving "critical collaborations" with mining companies—relationships that are "intimate and acrimonious, supportive as well as adversarial"—that can be instrumental in improving industry practice. Similarly, Bebbington writes that there are many brave actors who are in the business of building "alternative approaches to extraction that might leave more space for rights and environmental integrity—whether those elements are a land-use plan, a decision on Constitutional principles, or a moral sanctioning of a government department whose decisions have contravened civil rights" (Bebbington, 2012: 1161). Understanding these alternatives entails studying local culture, power dynamics, and the diversity of views likely to exist within a single community (Owen & Kemp, 2013) (see also Gilbert within this book).

The geologic turn in the social sciences draws readily from the work of STS-informed theorists, particularly surrounding the emergence of the Anthropocene (Latour, 2014; Mitchell, 2011). There is a rich tradition of STS work on energy infrastructure (Hughes, 1983; Nye, 1990; Winner, 1980). There is also an emerging interest in renewable energy ethics and policy, evidenced by the June 2013 special volume on Energy Transitions in *Science as Culture*. Given the importance of critical minerals to the material realities of a postcarbon economy, this is clearly a ripe area for further development. As Sujatha Raman argues, "the socio-technical networks of renewables extend beyond the point of energy supply to include materials, institutions, and publics involved in the production of these technologies." She goes on to write that a "socio-technical perspective requires making visible the political economy of metals required for renewable energy technologies" (Raman, 2013: 175).

An STS perspective on extraction examines the technoscientific aspects of how questions about mining are posed and deliberated, how extraction itself occurs, and how the consequences of such extraction are addressed (Kinchy, Phadke, & Smith, 2018). Underlying each of these areas are issues of knowledge, expertise, and power that STS is uniquely positioned to explore. My intervention in the critical minerals debate has been to connect the subsurface with the above ground and to consider how greenfield mining is one of many solutions for acquiring strategic minerals. In this way, engaged STS work on the underground also connects up with a lively interest in the field in addressing issues of environmental justice, climate justice, and toxic exposures. I am interested in how climate change solutions normalize the discourse around "critical" and "strategic" materials crises. Similarly, mining companies can maneuver frameworks around "corporate responsibility" and "social license to operate" with great interpretative flexibility. Most of the sustainable and responsible mining literatures emerge from transnational corporate coalitions, like the Mining Minerals Sustainable Development Project. Herein, "responsible mining" references the rectification of egregious environment and human rights conditions in far-flung places, like regions of Indonesia, Papua New Guinea, and southern Africa (Kirsh, 2014; Welker, 2014). The reemergence, or re-domestication, of critical minerals mining in the United States presents great challenges to the logic of "responsible" mining. Given the multinational scope of mining companies, "responsibility" and "social license to operate" may be insufficient analytical categories to serve the interests of the most vulnerable parties, because they have been too thoroughly monopolized by corporations and their political champions.

Methods

The study of critical minerals debates in the United States reported in this chapter is based on undergraduate faculty-student collaborative research. Over the last 3 years my research team has included undergraduate students enrolled in my environmental politics courses, and those I have hired for intensive summer research experiences. Our research has aimed to examine the discursive politics at work in the creation of a "strategic materials crisis" and in defining and deployment of "responsible" green-field mining as the solution. To better understand the potential to repatriate mining in the United States, my students and I have examined a series of case studies that represent varied geographic regions, landscape types, project scales, socioeconomic contexts, and corporate developers in eight states across the United States. Our goal has been to understand the places *where* projects are planned, *which* local actors engage in deliberative decision-making, *how* they engage in shaping project outcomes (organized public consultations, public hearings, and informal attempts), and *who* is disenfranchised or disengaged from the process. The specific questions we have addressed are:

- Where is the proposed mine and who owns the land?
- Which firms are developing the project and which minerals are being mined?
- Does this region have a history/legacy of mining?
- Which energy/mining and environmental impact rules are guiding development?
- Does the project have legislative/political support?
- How have local communities participated in the project?
- How are developers branding the project and has there been a good faith to address public concerns?

Our research team reviewed public documents, developer reports, and media accounts. We also conducted phone and e-mail interviews with key actors, including project developers, local journalists, and key community members involved in project opposition. Our most intensive research was focused on the Minnesota case, in the northern United States, where given our proximity to the region, we were able to conduct repeated fieldwork to visit proposed sites and interview local residents, business owners, tribal members, and environmental advocacy organizations.

Stories from the mining fields

The map given below documents the locations of the case studies we have examined. Our project website provides lengthier descriptions of each case.[a] I highlight three cases below that provide a view into the challenges of opening new mines, despite the federal government's interest in expediting the regulatory process, in Texas, Alaska, and Wyoming. Our Minnesota findings have been published at length elsewhere (Phadke, 2018). I chose to report on these three states in this chapter because they are the farthest along, are located in states deemed friendly to extractive industries, and sites where the investing firms might expect to get a great deal of leverage from an enthusiastic Trump Administration (Fig. 2.1).

Alaska

Located in the southeastern panhandle of Alaska on the Prince of Wales Island, the Canadian mining company Ucore Rare Metals has begun rare earth element mining operations at Bokan Mountain. This project aims to extract 5.3 million tons dysprosium, terbium, and yttrium from the largest rare earth deposit in the United States.

FIG. 2.1

Projects examined.

Credit: Ensia.

---

[a] Our project website provides lengthier descriptions of each case, including information about the precise metals being prospected and project status https://www.macalester.edu/miningfutures/stories.html.

Bokan Mountain is the largest known dysprosium deposit in US territory. This is an important mineral as it is estimated that Toyota will demand 10 tons of dysprosium per year on the Prius line alone (Fisher, 2013).

The project area ranges across 9500 acres of federal land in the Tongass National Forest. Ucore reports that the mine would employ about 300 people during the 2-year construction period and would create almost 200 jobs during its 10 years of operation. These economic promises have resulted in widespread support from local politicians for this project. In Alaska, tax revenues from mining operations provide payroll for state workers and educational funding. In 2014, former Governor Sean Parnell signed legislation authorizing up to $145 million in public financing for the project, which is nearly three-quarters of its $200 million price tag (Brehmer, 2016).

Expedited review processes have many native tribes concerned. The Haida Indian Tribe, the Ketchikan Indian Community, and the Craig Tribal Association were all consulted during the initial environmental assessment conducted by the US Forest Service. Many of them have been wary of the possible impacts that mining will have on their traditional subsistence practices, especially given the fact that Bokan Mountain is already listed as a Superfund site contaminated by previous uranium mining (Public Radio International, 2013). Moreover, there has been concern from multiple parties that the jobs created by the Bokan Mountain Project would go to nonlocals and not actually help revitalize the economy of the area. The Tongass Conservation Society in southeastern Alaska is worried that Ucore is downplaying the potential damage to the Kendrick Creek, critical habitat for salmon populations that are harvested by locals for subsistence.

In 2013, the US record of decision was released by the US Forest Service for Ucore's Mining Plan of Operations. It stated that the project would not have any adverse environmental impacts, provided that some modifications were put in place. In early 2018, Ucore announced plans to open a rare earth metals separation plant in nearby Ketchikan. Although construction of the Bokan Mountain mine itself has not begun, the company eventually plans to process the ores from the mine at the plant along with ores from all over the world. More recently, in 2019, Alaska's current Governor, Michael Dunleavy, asked the Trump White House to designate Bokan as a "High Priority Infrastructure Project." This designation would further fast track the environmental review process.

Of all the critical mineral deposits being currently prospected in the United States, Bokan seems like closest to fruition. Yet, it is still nearly a decade away from producing and processing those minerals. There is a high likelihood that environmental and indigenous rights organizations will sue the state and the federal government over the landscape and cultural impacts.

Wyoming

In comparison to Bokan, the Bear Lodge project is limping along. Located in northeastern Wyoming, 20 miles from the South Dakota border and 22 miles southeast of the famous Devil's Tower, Rare Element Resources Ltd. has proposed a rare earth elements mine named Bear Lodge. This area, situated in the Bear Lodge Mountains on the unique Black Hills Uplift, contains one of the largest deposits of rare earth elements (REEs) in North America according to the US Geological Survey.

The proposed Bear Lodge mine would mine 10 of the 17 rare earth elements including neodymium, dysprosium, and europium. Covering approximately 1700 acres in the Black Hills National Forest, the proposed Bear Lodge mine has a predicted lifetime of 43 years. According to a 2014 prefeasibility study released by the company, this $290 million open-pit mine would extract 423,000 tons of ore annually when operating at full capacity (RER, 2014).

Exploration and drilling for the Bear Lodge mine began in 2011. In January 2016, Rare Element Resources suspended all permitting processes due to low market prices and a lack of funding from investors. This happened shortly after the US Forest Service released a Draft Environmental Impact Statement that recommended the advancement of the project (Fladager, 2016). Activity resumed again when the San Diego-based, defense-contracting company, General Atomics invested $4.7 million into the company in 2017.

The state of Wyoming has an extensive mining history and is generally considered friendly to extractive industries. The state supplies more energy in the form of fossil fuels to the country than any other state and is the nation's leader in coal production (IER). Historically, coal mining in Wyoming has been the cornerstone of the state's economy. The coal industry in nearby Gillette has suffered recent job loss with declines in production. The jobs created by the Bear Lodge Project could contribute to a revitalization of the area's economy.

Significant concerns have been expressed by many groups in the area about this project. The most vocal input occurred during a public meeting held by the Black Hills National Forest where around 140 people attended. A majority of the concerns were about the potential for wastewater contamination and high levels of expected dust (Peterson, 2014). In addition, Bear Lodge's proximity to Devil's Tower posed many concerns for local native tribes due to the cultural and historical significance that the unique geologic feature holds for them. Devils Tower, or Mahto Tipila, holds cultural significance for more than 20 tribes in the area who visit the site for personal and group healing and spiritual rituals. In response to the proposed development, the grassroots, nonprofit organization, Defenders of the Black Hills, organized two prayer gatherings at the foot of Devils Tower to protest impacts to sacred lands and burial sites in the area (Nauman, 2016). When given the opportunity to respond to public concerns about the mine, Rare Element Resources did not address the concerns posed by native communities.

Support for the Bear Lodge Mining project rests of the job creation interests of local community residents and the mineral needs of the national government. Despite this, some consider the mine to be too small to meaningful revolutionize the global market (Storrow, 2014). Additionally, the mine is still subject to the investor whims of the global market. To date, the processing plant is near completion. It may be another decade before the mine is open for production.

Texas

The Round Top Mountain mine is a $350 million project being developed by a partnership between Texas Mineral Resources and USA Rare Earth. The project is in the Sierra Blanca Mountain Range of western Texas in Hudspeth County, not far from the border town of El Paso. This company seeks to mine yttrium and dysprosium while also extracting uranium, beryllium, and thorium. According to the company website, they hold 19-year renewable leases from the State of Texas on 950 acres encompassing Round Top, and additional permits on adjacent lands covering more than 9000 acres. The project developers claim they will create up to 150 new jobs in the area (Hatch, 2014). Moreover, the company will be required to pay a 6.25% net smelter royalty to the state which will be used mainly to fund Texas' educational system.

As of 2016, an estimated 4053 people lived in Hudspeth County of which 77.6% of the population identified as Hispanic or Latino and the median household income was $35,252. The county is quite poor with 21.4% of residents living below the poverty line, which is 6% higher than the national average.[b] Mining began in the area in the 1940s with the discovery of zinc. Since then coal, silver, and copper have been extracted as well (Kohout, 2019). Due to the decline in the local agriculture and mining economies in recent years, there is a common desire for an industry that could help support the region. Until recently, agriculture and ranching also supported the economy of Hudspeth County.

The Round Top Mountain Project is operating within a complex and unique policy context. Perhaps the most notable aspect of the proposed mine is that the lands are owned by the state's General Land Office (GLO), not federally operated institutions such as the Bureau of Land Management or the Forest Service as is often the case with proposed mines. Because the lands are owned by the state, the Round Top Mountain Project will not trigger the National Environmental Policy Act (NEPA) and thus escape lengthy federal environmental review. Texas is known for some of the most extraction-friendly regulations in the country and the GLO has little oversight from other regulatory agencies. The company will need to acquire 12 local permits from Hudspeth County.

While the CEO of USA Rare Earths recently argued that the mine is targeted for full production in 2 years, there has been concern from the public concerns may stall the project. In particular, surrounding communities, such as Sierra Blanca, are apprehensive about potential health risks from radioactive substances such as uranium, thorium, and beryllium. Other concerns include excessive dust exposure as a result of the mine, which can cause detrimental health effects like cancer and berylliosis. Others are anxious that the sulfuric acid solution from the heap leaching process may leak into surrounding groundwater and contaminate water supplies in the county. Finally, some worry that the mine will destroy the mountains that the county is known for, and that the jobs created by the mine will not actually be given to local residents (Hudspeth County Herald, 2014).

Some of these concerns predate this mining project. Sierra Blanca residents have been concerned about environmental justice issues for two decades. The Latinx community in the area has expressed frustration with the large amount of hazardous materials, including biosolids from New York City's sewage that have been proposed for deposit near the community and put them at risk. Area residents fought off the siting of a nuclear waste dump in their region 20 years ago. The TMRC mine seems to some to continue these trends of exposure (Lyman, 1998). Most of the town's employment comes from working 90 miles away at the Border Patrol checkpoint.

---

[b] This data was retrieved from Hudspeth county from https://datausa.io/profile/geo/hudspeth-county-tx.

While Texas Mineral Resource Corporation seems to have the general support of the state, the TMRC project will still operate on the whims of the global market, which has been the largest obstacle to the advancement of the mine. This puts the viability of the entire project into question.

Critical elements and social license

Looking across these three cases, none of these projects is near commercial production. Yet, each project has been advanced by project developers as antidote to a rising trade war with China and the need for domestic supplies of critical elements. Their success now depends in big part on the Trump Administration's willingness to subsidize the production and expedite regulatory review.

Analysis of these cases reveals several interesting insights about the prospects for new critical minerals mining in the United States. First, a "social license" to mine in the contemporary period is driven by historical mining identities and party politics, as much as by purely business/economic rationales. Yet, even in places with a historical legacy of mining, social opposition is present from native populations and minority groups. For example, in the above Texas case, environmental justice concerns have been raised by Latinx communities who contest the health impacts and a legacy of dumping. Second, global commodity price shifts have significant impacts on project development (see Gilbert within this book). Proposed mines we examined in Texas and Wyoming that passed the permitting phase, failed to meet their timetables because companies could not raise enough money on capital markets. Third, environmental review is minimal in many states, especially those with pro-mining/extraction policies like Texas (see McLellan within this book). Herein, communities and tribes will lean on national networks to support their protest efforts. Considering the national and international attention that Standing Rock protests received, it is not unthinkable that mining controversies could similarly explode. The challenge is that clean energy advocates will find themselves backed into a corner as they weigh local environmental injustices with global climate struggles.

Despite claims of urgency and the need for expedited review, the Trump Administration has found it exceedingly difficult to launch a new era of rare earths mining in the United States. As Hefferman (2015) noted in *High Country News*, "a new American mining boom is nowhere in the cards, and the global nature of ECE production means that neither Silicon Valley nor Washington, DC, can do much to change that."[c] Even in the most mining-friendly parts of the nation, new mines need immense capital investment, face volatile global commodity markets and experience limited host community support.

This raises important questions about what role the federal government should play in supporting new critical minerals development. The US Congress has focused on passing legislation, like the 2017 METALS Act (H.R. 1407), to provide incentives

---

[c] ECE here refers to energy critical elements. The terms green tech metals and climate action metals are also used to describe this same subset of elements.

to industry to develop new sites. Given the patchwork nature of the environmental review and mounting opposition, the federal could play a larger role in providing criteria for siting and permitting. Federal incentives could also be aimed at state agencies to support decision-making and greater public engagement.

An important blind spot in these debates is the near-complete focus on greenfield mining for supplies of critical minerals. There seems to be a shocking lack of attention to opportunities that come from investing in metals recovery and recycling (see Godding, Chapter 12). Of course, there are significant challenges to recovering critical minerals. This includes a lack of recycling infrastructure for electric vehicles and solar panels. Research programs at the US Department of Energy's Ames and Oak Ridge National Labs have begun to investigate how to process spent critical metals for reuse. Yet, the national conversation has been dominated by a discourse of metals scarcity rather than a lack of responsible sourcing.

In contrast, consumer technology corporations are moving headlong into this field, knowing well that their customers have already begun demanding products designed with responsibly sourced minerals, such as conflict-free cobalt and lithium (see Franken et al. and Prause within this book). For example, Apple recently announced that it aims to source 100% of the minerals used in iPhones and Macbooks from conflict-free, recycled, and renewable supply chains (Apple, 2019).

An emerging social science research agenda

Given the geopolitics of critical minerals and the current development imperative for new domestic sources, this is an important moment for policy makers and citizen groups to ask a range of descriptive and normative questions about how "sustainable" and "responsible" mining can be defined, practiced, and regulated. Social scientists have an important role to play in investigating if and how critical minerals development will gain broad-based public acceptance. Balancing long-term climate mitigation and adaptation priorities with local community environmental justice concerns is a significant challenge. As new mining projects touch down in local communities, it is not clear that the willingness for such a bargain exists, particularly when these processes are led by state level agencies in extractive friendly states. The staunch opposition to new mining has been driven by environmental justice activists who call out the unfair ecosystem and body burden that will be borne by those who live near mining sites, particularly native groups. This contamination will last generations.

The "responsible" mining discourse also further naturalizes the claim that climate change action rests on solving the "strategic materials crisis" with silver bullet technologies. In other words, climate solutions rest in making and consuming more of the smarter, greener technologies we demand from raw materials. This leaves out the important role recovery and recycling play in sourcing metals. For example, urban mining is an emergent field at the intersection of waste management and mining studies. The phrase refers to the practice of extracting resources from electronics and other anthropogenic sources of waste for their recycling and reuse in infrastructure and technology. Scholars need to be asking what kinds of planning, business incentives, and infrastructures will be necessary to shift attention toward secondary mining. As a corollary, there is also much to be learned about the kinds of social dislocation and community opposition that have occurred near urban mining centers, such as the case of Umicore's Solvay plant in France.

Another important issue is the role of labor transitions in the sourcing of sustainable minerals. In the US cases we have examined, struggling rural communities are drawn to the promise of new high-wage jobs. Just transition platforms foreground the need to link climate solutions to building strong local economies that provide well-being for all. This particularly applies to those currently employed in extractive industries. A social science agenda on the future of critical minerals can engage with climate policies, like the Green New Deal in the United States, to carve out opportunities for climate and economic justice that focus on recycling and recovery markets.

References

Apple. (2019). *Material impact profiles*. Available from: https://www.apple.com/environment/pdf/Apple_Material_Impact_Profiles_April2019.pdf.

Bebbington, A. (2012). Underground political ecologies. *Geoforum, 43*(6), 1152–1162.

Benson, P., & Kirsch, S. (2010). Corporate oxymorons. *Dialectical Anthropology, 34*, 45–48.

Brehmer, E. (2016). Bokan mine development slowed as rare earth prices dip. *Alaska Journal of Commerce*. Available from: http://www.alaskajournal.com/2016-04-20/bokan-mine-development-slowed-rare-earthprices-dip#.WouAyuinG3B.

Centers for Disease Control (CDC). (2018). *Prevalence of black lung continues to increase among U.S. Coal Miners*. Press release. Available from: https://www.cdc.gov/niosh/updates/upd-07-20-18.html.

Clagett, N. (2013). A rare opportunity: streamlining permitting for rare earth materials within the United States. *Journal of Energy & Environmental Law*, (Summer), 123–138.

Clark, N. (2011). *Inhuman Nature: Sociable Life on a Dynamic Planet*. London: Sage.

Fisher, S. (2013). *Will a small Alaskan town be forced to trade its ecosystem for mining jobs*. In These Times. Available from: http://inthesetimes.com/article/15260/in_alaska_rare_earth_discovery_pits_jobs_against_environment.

Fladager, G. (2016). Proposed rare earth mine suspended indefinitely. *Casper Star Tribune*. 25 January 2016. Available from: http://trib.com/news/local/casper/proposed-rare-earth-mine-suspended-indefinitely/article_cb93d6ba4c33-5f9b-80e3-fd716a5bfb72.html.

Friedman, L. (2019). What is the green new deal? A climate proposal, explained. *NY Times*. 21 February 2019. Available from: https://www.nytimes.com/2019/02/21/climate/green-new-deal-questions-answers.html.

Goodland, R. (2012). Responsible mining: The key to profitable resource development. *Sustainability, 4*(9), 2099–2126.

Hatch, G. (2014). A visit to the Texas rare earth resources round top project. *Technology Metals Research*. 19 May 2014. Available from: http://www.techmetalsresearch.com/a-visit-to-the-texasrare-earth-resources-round-top-project/.

Hefferman, T. (2015). Why rare-earth mining in the West is a bust. *High Country News*. 16 June 2015. Available from: https://www.hcn.org/issues/47.11/why-rare-earth-mining-in-the-west-is-a-bust.

Hudspeth County Herald. (2014). *Public airs concern about round top mine.* 31 January 2014. Available from: http://hudspethcountyherald.com/public-airs-concerns-about-round-top-mine/.

Hughes, T. (1983). *Networks of Power: Electrification in Western Societies.* Baltimore: John Hopkins Press.

Kinchy, A., Phadke, R., & Smith, J. (2018). Engaging the underground: An STS field in formation. *Engaging STS, Vol. 4.* Available from: https://estsjournal.org/index.php/ests/article/view/213.

Kirsh, S. (2014). *Mining capitalism: The relationship between corporations and their critics.* Berkeley: UC Press.

Kohout, M. (2019). *Hudspeth County. Texas State Historical Association.* Available from: https://tshaonline.org/handbook/online/articles/hch21.

Latour, B. (2014). Agency at the time of the Anthropocene. *New Literary History, 45,* 1–18.

Li, F. (2011). Engineering responsibility: Environmental mitigation and the limits of commensuration in a Chilean mining project. *Focaal—Journal of Global and Historical Anthropology, 60,* 61–73.

Lyman, R. (1998). For Some, Texas town is too popular as waste disposal site. *New York Times.* 2 September 1998. Available from: http://www.nytimes.com/1998/09/02/us/for-some-texas-town-is-toopopular-as-waste-disposal-site.html.

McAllister, M., Fitzpatrick, P., & Fonseca, A. (2014). Challenges of space and place for corporate 'citizens' and healthy mining communities: The case of Logan Lake, BC and Highland Valley Copper. *The Extractive Industries and Society, 1*(2), 312–320.

Mitchell, T. (2011). *Carbon democracy.* London: Verso Press.

Nauman, T. (2016). Canadians suspend bid to mine near Devil's Tower. *Native Sun News.* 27 January 2016. Available from: https://popularresistance.org/canadians-suspend-bid-to-mine-near-devils-tower/.

Nye, D. (1990). *Electrifying America: Social Meanings of New Technology.* Cambridge: MIT Press.

Owen, J. R., & Kemp, D. (2013). Social licence and mining: A critical perspective. *Resources Policy, 3*(8), 29–35.

Peterson, C. (2014). Residents question safety of proposed rare-earth mine. *Billings Gazette.* 16 April 2014. Available from: http://billingsgazette.com/news/state-and-regional/wyoming/residents-question-safety-of-proposed-rareearth-mine/article_7f796903-1063-5c50-8168-e1e498a0b55f.html.

Phadke, R. (2018). Green energy futures: Responsible mining on Minnesota's iron range. *Energy Research & Social Science, 35.*

Public Radio International. (2013). Alaska tribal community worries about pending rare earth element mine. *PRI.* 10 July 2013. Available from: https://www.pri.org/stories/2013-07-10/alaska-tribal-community-worries-about-pending-rare-earth-element-mine.

Raman, S. (2013). Fossilizing renewable energies. *Science as Culture, 22*(2), 172–180.

RER. (2014). *Rare Element Resources Files NI 43-101 Technical Report on Positive Pre-feasibility Results for Bear Lodge Project.* Business Wire. 10 October 2014. https://www.goldseiten.de/artikel/221935--Rare-ElementResources-Files-NI-43-101-Technical-Report-on-Positive-Pre-feasibility-Results-for-Bear-LodgeProject.html.

Ross, M. R. V., McGlynn, B. L., & Bernhardt, E. S. (2016). Deep impact: Effects of mountaintop mining on surface topography, bedrock structure, and downstream waters. *Environmental Science & Technology, 504,* 2064–2074.

Scheyder, E. (2019). *Exclusive: Pentagon races to track U.S. rare earths output amid China trade dispute*. Reuters. Available from: https://www.reuters.com/article/us-usa-rareearths-pentagon-exclusive/exclusive-pentagon-races-to-track-us-rare-earths-output-amid-china-trade-dispute-idUSKCN1U727N.

Sierra Club. (2019). *Ready for 100 campaign*. Available from: https://www.sierraclub.org/ready-for-100.

Smith Rolston, J. (2015). Turning Protesters into Monitors: Appraising critical collaboration in the mining industry. *Society & Natural Resources, 28*(2), 165–179.

Storrow, B. (2014). *Is a Black Hills rare earth min in America's interest? Casper Star Tribune.* 27 September 2014. Available from: http://trib.com/business/energy/is-a-black-hills-rare-earth-mine-in-america-s/article_4eb5665e-3fbb5090-b6a8-bfd8a433eeea.html.

Welker, M. (2014). *Enacting the Corporation: An American Mining Firm in Post-Authoritarian Indonesia*. Berkeley: UC Press.

Winner, L. (1980). Do artifacts have politics? *Daedalus, 109*(1), 121–136.

Yusoff, K. (2013). Geologic life: prehistory, climate, futures in the anthropocene. *Environment and Planning D: Society and Space, 31*(5), 779–795.

Metal-energy nexus in the global energy transition calls for cooperative actions

3

Peng Wang[a], Nan Li[a], JiaShuo Li[b], and Wei-Qiang Chen[a]

[a]*Institute of Urban Environment, Chinese Academy of Sciences, Xiamen, China*
[b]*Institute of Blue and Green Development, Shandong University, Weihai, China*

Introduction

Energy is essential to human civilization. It is harvested, processed, and stored through a series of energy infrastructures such as power plants, high-voltage transmission lines, petroleum refinery plants, batteries, etc. To stabilize our climate, the global energy system is on the verge of a major shift from fuel-based energy to renewables. This poses various enormous challenges. In particular, the development of renewables to such high levels requires a substantial amount of diverse minerals (WWF, 2014), and most of those minerals are not only essential, but also extremely scarce and have high levels of risk involved in their global supply chains. Given that renewable energy systems are more dependent on the critical materials such as rare-earth elements and gallium than traditional fossil fuel-based systems (American Physical Society, 2011), there is an urgent need to incorporate mineral constraints into future energy planning as well as climate policy making.

At present, the primary focus of the energy system transition is still on issues such as cost, the environment, policy mechanisms, etc. Information about critical materials related to emerging renewable energy technologies is quite rare. Recently, the study of critical materials (or minerals, hereafter referred to as critical metals) has been increasing. For instance, the US National Research Council (NRC) has identified the risks involved in the supply of various critical metals (NRC, 2008), and the US Department of Energy (DOE) has issued two critical materials strategy reports, in 2010 and 2011, respectively (U.S. DoE, 2010, 2011). Meanwhile, the European Commission has also published three reports investigating critical materials of great importance to the European Union (EU), in 2010 (EU Commission, 2010), 2014 (EU Commission, 2014), and 2017 (EU Commission, 2017), respectively, and most of those materials related to renewable energy technologies. Other nations, such as Australia (Skirrow et al., 2013) and Japan (Hatayama & Tahara, 2015), have also expressed strong concerns regarding the risks involved in the supply of critical materials. The combination of metal criticality studies and energy systems (from

The Material Basis of Energy Transitions. https://doi.org/10.1016/B978-0-12-819534-5.00003-9

generation, transmission, and storage to consumption) can help to offer a holistic view of this ongoing global energy system transition.

Hence, this chapter aims to introduce the context, framework, implementation, and implications of metal-energy nexus thinking, and to unite the domains of material systems and energy system studies for the successful transition of the global energy system. The chapter consists of three sections: the first section introduces the background of this study by tracing the global energy system transition from 1800 to 2050, thereby highlighting that the global energy system is shifting from carbon driven to metal driven. The second section presents the metal-energy framework of the metal-energy nexus, and uses this framework to analyze 18 types of energy-critical materials from a material cycle perspective. The final section explores the potential metal constraints, and urges all stakeholders throughout the material cycle and trade network to take cooperative action to promote the sustainable metal and energy transition.

Global energy system shifts from carbon driven to metal driven

This section will first introduce the evolution of the global energy system from 1800 to 2050, which highlights the growing importance of the metal-energy nexus on future energy system planning. Accordingly, the link between energy and critical metals will be described in the subsection about energy-critical elements.

Energy transition: Becoming more metal driven?

Energy is essential for human well-being and national prosperity, thus the evolution of human societies is closely linked to energy transitions (Smil, 2010). Fig. 3.1 highlights two of the three main features of past energy transitions: the exponential growth in global energy use, and the shift from biofuels and fossil fuels to renewable energy sources. The third feature is the evolution from carbon to metals, which is explained in detail below.

(1) **Energy demand: Exponential growth in global energy use**. Global energy use has witnessed exponential growth in the past 200 years, mainly after the Industrial Revolution. The amount and type of energy use between 1800 and 2017 is illustrated in Fig. 3.1A in terawatt hours (TWh) per year. In 2015, the world consumed 146,000 TWh of primary energy—more than 25 times the amount consumed in 1800. During the second half of the 20th century, energy consumption increased sharply around the same time as the energy transition from coal to oil. The world's total energy requirements have tripled during the past 50 years. After the World War II, the energy consumption kept accelerating with the amount of oil in the energy mix increased rapidly to 40% by the 1970s.

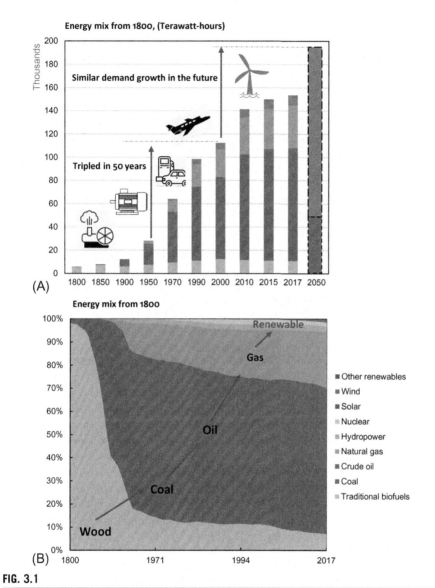

FIG. 3.1

The growth of global energy use (A) and the evolution of energy sources (B) overtime.

Credit: Compiled by the author on the basis of data from Ritchie, H., Roser, M. (2018). Energy production & changing energy sources [WWW document]. Our world in data. https://ourworldindata.org/energy-production- and-changing-energy-sources (Accessed 9 January 2019).

Looking to the future, total energy supply will continue to increase to meet the unprecedented demand worldwide that is forecast by various institutions (IEA, 2018; IRENA, 2019). For instance, the International Energy Agency (IEA) predicts that global energy demand will grow by more than a quarter between 2017 and 2040. Meanwhile, the deep decarbonization of our energy system is becoming increasingly urgent for the stabilization of our climate (Davis et al., 2018). The amount of renewable energy—particularly wind and solar—in the energy mix has rapidly increased in the past decade, and renewables growth is faster than any other energy source. In 2017, the global share of renewables (excluding large-scale hydro) had risen to 61% of newly added power generation capacity (IRENA, 2017).

The future energy transition will also be driven by the developing nations such as China and India. It is particularly worth noting that China has become the world's largest energy consumer, and its usage is expected to increase to around 1590 petajoules (PJ) by 2050. As the world's largest energy consumer and carbon emitter, China has made the development of renewables one of its national priorities and is galloping ahead in the global race for renewable investment, manufacturing, employment, and power generation (Wang, Chen, Ge, Cai, & Chen, 2019). From 2035 to 2050, around 60% of the country's primary energy demand will be met by renewables, and this figure could even reach 100% in certain areas (KAPSARC, 2018). Meanwhile, China is also leading the transition toward electric transportation: by 2030, it is expected that more than a quarter of the Chinese car market will consist of electric vehicles (IEA, 2019).

(2) Energy source: From biofuel, coal, and oil to renewables. The sources of energy have also changed dramatically, evolving from biofuel, coal, and oil (gas) to renewables, as shown in Fig. 3.1B. Technology plays a critical role in such source transitions. Primitive humans began burning wood and other biofuels for heating, cooking, and other basic needs. With the development of exploration and mining technologies, coal surpassed wood and became the main energy source from the 1780s onwards. The invention of the steam engine by James Watt in 1769 accelerated the use of coal for industrial purposes, and the world's first coal-fired power plant was built in France in 1875. In 1886, the internal combustion engine was invented by Daimler, which accelerated the consumption of oil and gas for daily transportation purposes. The global energy mix changed accordingly. Before 1900, the share of biofuel in the primary energy mix was around 51%, followed by coal with 47%. Then in 1970, the share of oil and gas rose rapidly to 57%. Since the start of the 21st century, there has been an ongoing transition from fuels to renewables, but clean energy still makes up less than 6% of the global primary energy mix. In order to meet the targets of the 2015 Paris Agreement, IRENA (2019) predicts renewables will have to account for 85% of global power generation, and solar and wind energy will need to become the main sources of power by 2050. This energy transition is also being shaped by the increasing electrification of transportation worldwide. Between 2015 and 2018, annual sales of electric passenger cars increased by nearly 60%, and by the end of that period the global fleet of electric vehicles exceeded 5.1 million (IEA, 2019). According to the IEA's New Policy Scenario (2019), by 2030 global EV sales will reach 23 million and the fleet will exceed 130 million.

(3) **Element basis: From carbon to metals**. Energy technologies, including conventional and novel infrastructure for generating, storing, transmitting, and conserving energy, are enabled by various specific materials. During the course of the energy transition, its nexus with metals (or metalloids) has changed dramatically, shifting from carbon-driven period to metal-driven period. In the 1700s, most energy-related tools were built from stone, wood, and iron. These materials consist of three main elements: carbon (C), calcium (Ca), and ferrum (Fe). During the industrial revolution when coal and gas became the major sources of energy, most boilers, steam engines, and pipes were made of steel, copper, and their alloying elements (e.g., Cr, Mn, and W). In the 20th century, the spread of the internal combustion engine prompted the shift from coal to oil, and using oil to generate energy necessitates specialized steels with a range of alloying elements, including Cr, Ni, Mo, Mn, Co, V, and W. Producing and refining petroleum requires catalysts that contain elements such as Pt, Rh, rare-earth elements (REEs), Re, Pd, Mo, and V. Moreover, Cu is an essential element, especially in the automotive industry. The push toward decarbonization and the associated rise of renewable energy technologies is causing the shift toward a metal-driven period of energy generation. Photovoltaics (PV) panels require the elements Cd, GaN, Ge, and Te. Some wind turbines need REEs to improve the efficiency of their super strong magnets (see Goddin, Chapter 13). Lithium, cobalt, nickel, and lanthanum are essential for high-performance batteries, which will be widely used in electric vehicles, and nuclear energy requires uranium and a range of materials as moderators. At present, more than 60 metallic elements are involved in energy pathways (Zepf, Simmons, Reller, Ashfield, & Rennie, 2014).

Energy-critical elements: Nexus energy and critical metals

'Critical metals' (minerals or other materials) refer to a group of metals such as rare-earth elements, cobalt, and indium, which are vital for the emerging technologies of the global sustainable energy transition (Graedel, Harper, Nassar, Nuss, & Reck, 2015). Metal(Material) criticality in itself has attracted a lot of attention worldwide, including various review papers (Frenzel, Kullik, Reuter, & Gutzmer, 2017; Graedel et al., 2012; Graedel & Reck, 2016; Habib & Wenzel, 2016; Hayes & Mccullough, 2018; Ioannidou, Heeren, Sonnemann, & Habert, 2019; Jin, Kim, & Guillaume, 2016; Løvik, Hagelüken, & Wäger, 2018). In 2008, the US National Research Council (NRC) (2008) published the first report to screen critical materials from various material candidates by measuring material criticality in terms of two-dimensional (2D) indicators: supply risks and the impact of supply restriction (see also Gilbert, Zepf, McLellan, and Koch within this book). A similar framework is deployed by the EU in its critical raw material assessments (2017, 2014, 2010), as well as by General Electric (Duclos, Otto, & Konitzer, 2009), Volkswagen (Rosenau-Tornow, Buchholz, Riemann, & Wagner, 2009), UNEP (UNEP, 2009), the United States Department of Energy (US DoE, 2010), and others.

Critical metals that are closely linked to energy technologies are referred to as "energy-critical elements" (ECEs). This term was proposed by a joint committee of the American Physical Society (APS) and the Materials Research Society (MRS) in its report "Securing Materials for Emerging Technologies" in 2009 (American Physical Society, 2011). This report proposes a definitive list of ECEs according to the information about occurrences, reserves, extraction, processing, utilization, and recycling. The potential ECE candidates include (a) gallium, germanium, indium, selenium, silver, and tellurium in advanced photovoltaic solar cells; (b) dysprosium, neodymium, praseodymium, samarium (all REEs), and cobalt in wind turbines and hybrid automobiles; (c) lithium and lanthanum in high-performance batteries; (d) helium in cryogenics and nuclear applications; (e) cerium, platinum, palladium, and other platinum group elements (PGEs) in fuel cells as catalysts; and (f) rhenium in high-performance alloys for advanced turbines. Notably, various metals like copper, aluminum, iron, tin, and nickel are excluded due to their large, mature, and vigorous markets with low supply risks. By combing key reports from BP (Zepf et al., 2014), WWF (WWF, 2014), the US Department of Energy (US DoE, 2010), and ITO (Rietveld, Boonman, van Harmelen, Hauck, & Bastein, 2019), we mapped the energy-critical elements with their corresponding applications in Fig. 3.2.

The sustainable material cycle in the global energy transition

The increasing interdependence between energy and metals calls for an urgent and detailed analysis of the energy-critical metals that underpin a successful global energy transition. We have proposed a metal-energy nexus framework that shows the synergy of metal systems and energy systems for a systematic analysis (cf. Wang et al., 2019; Wang & Kara, 2019). For a detailed analysis of material cycles, we apply a technique called material flow analysis (MFA) to quantitatively characterize the material flows and stocks along the life cycle of different materials (Chen & Graedel, 2012). In the past decade, the material cycles of several critical metals have been analyzed using the MFA approach, including indium (Licht, Peiró, & Villalba, 2015), gallium (Løvik, Restrepo, & Müller, 2015), tellurium (Kavlak & Graedel, 2013a), selenium (Kavlak & Graedel, 2013b), and rare earths (Du & Graedel, 2011). This material cycle quantification not only provides thorough information about the stocks and flows in each lifecycle stage. More importantly, it can provide a more accurate estimation of metal demand by taking the various dissipative resource losses into account.

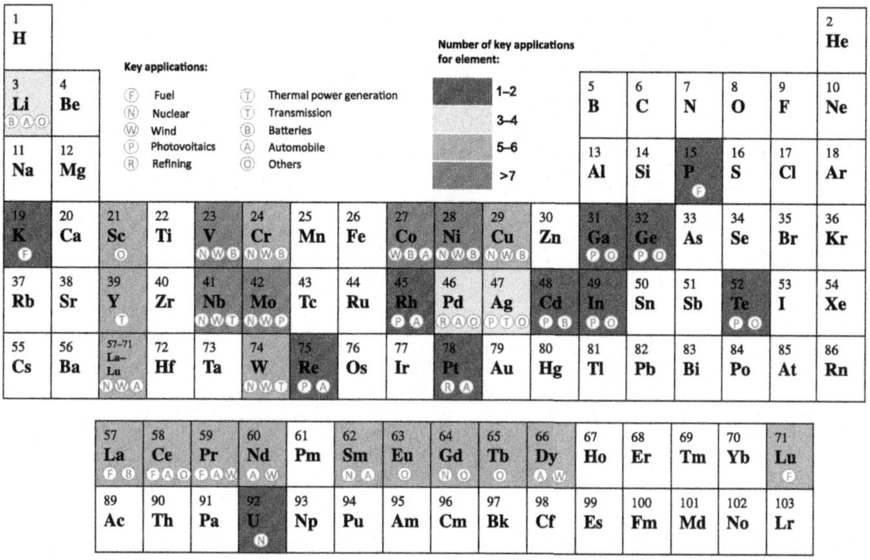

Key applications:

- (F) Fuel
- (N) Nuclear
- (W) Wind
- (P) Photovoltaics
- (R) Refining
- (T) Thermal power generation
- (T) Transmission
- (B) Batteries
- (A) Automobile
- (O) Others

Number of key applications for element:

- 1–2
- 3–4
- 5–6
- >7

Periodic table of energy-critical elements:

1 H																	2 He
3 Li (B A O)	4 Be											5 B	6 C	7 N	8 O	9 F	10 Ne
11 Na	12 Mg											13 Al	14 Si	15 P (F)	16 S	17 Cl	18 Ar
19 K (F)	20 Ca	21 Sc (O)	22 Ti	23 V (N W B)	24 Cr (N W B)	25 Mn	26 Fe	27 Co (W B A)	28 Ni (N W B)	29 Cu (N W B)	30 Zn	31 Ga (P O)	32 Ge (P O)	33 As	34 Se	35 Br	36 Kr
37 Rb	38 Sr	39 Y (T)	40 Zr	41 Nb (N W T)	42 Mo (N W P)	43 Tc	44 Ru	45 Rh (P A)	46 Pd (R A O)	47 Ag (P T O)	48 Cd (P B)	49 In (P O)	50 Sn	51 Sb	52 Te (P O)	53 I	54 Xe
55 Cs	56 Ba	57-71 La-Lu (N W A)	72 Hf	73 Ta	74 W (N W T)	75 Re (P A)	76 Os	77 Ir	78 Pt (R A)	79 Au	80 Hg	81 Tl	82 Pb	83 Bi	84 Po	85 At	86 Rn

57 La (F B)	58 Ce (F A O)	59 Pr (F A W)	60 Nd (A W)	61 Pm	62 Sm (N A)	63 Eu (O)	64 Gd (N O)	65 Tb (O)	66 Dy (A W)	67 Ho	68 Er	69 Tm	70 Yb	71 Lu (F)
89 Ac	90 Th	91 Pa	92 U (N)	93 Np	94 Pu	95 Am	96 Cm	97 Bk	98 Cf	99 Es	100 Fm	101 Md	102 No	103 Lr

FIG. 3.2

Energy-critical elements and their corresponding major applications.

Credit: Compiled by the author using data from BP (Zepf, V., Simmons, J., Reller, A., Ashfield, M., Rennie, C. (2014). Materials critical to the energy industry. An introduction (2nd ed.). London, United Kingdom: BP Publisher), WWF (WWF (2014). Critical materials for the transition to a 100% sustainable energy future. Gland, Switzerland: World Wide Fund for Nature), US DoE (U.S. DoE (2010). Critical materials strategy 2010. Washington DC: United States Department of Energy), and ITO (Rietveld, E., Boonman, H., van Harmelen, T., Hauck, M., Bastein, T. (2019). Global energy transition and metal demand—An introduction and circular economy perspectives. The Hague, Netherlands: TNO, Netherlands Organisation for Applied Scientific Research).

Here, we propose a quantitative framework for the analysis of energy-critical metals in Fig. 3.3. In the center, we deployed dynamic MFA to map the historical material cycles of energy-critical metals at the global level (cf. Wang, 2019). This involved studying various processes and activities along the metals' cycles, from material production (or production), manufacturing (or fabrication), to the in-use and end-of-life stages. Furthermore, those stages can be divided into several substages if greater detail is required.

We also complied "indicators of concern" to draw attention to potential constraints on the future supply and demand of those critical metals in energy-related applications based on a series of previous work (Graedel et al., 2015; Zepf et al., 2014). We selected six key indicators for this purpose, including RY (reserve-to-production year) to express the physical scarcity, PY (production yield rate) to show resource loss in mining and processing stages, EI (environmental impacts) to show the severity of impact from those production activities to ecosystems and human health, Trade to express the necessity to balance the demand and supply and associated supply risk from producers, SU (substitutability) to show the difficulty of substituting a given

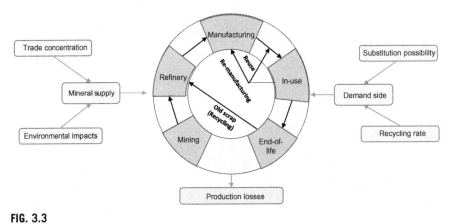

FIG. 3.3

Framework for a system analysis of energy-critical metals.

Credit: Compiled by the author.

material for a particular application, and RE (recycling rate) to show the level of recycling performance. In total, there are around 18 types of energy-critical materials under investigation in this chapter, which are divided into four major groups according to their key applications: battery technology, photovoltaic technology, wind or motor-related technology, and vehicle-related (e.g., automotive catalyst) technology.

(1) **Battery-related metals**. The globally accelerating electrification of the automotive sector will cause unprecedented demand for battery-related metals, including lithium, cobalt, nickel, vanadium, etc. (see Weil et al. within this book). The historical global cycle of those selected materials and their indicators of concern is illustrated in Fig. 3.4. Cobalt and nickel are relatively scarce with an RY of less than 68 and 36 years, respectively, while lithium and vanadium reserves are rich enough to meet future demand. All metals can cause high environmental impacts during their extraction and processing stages, which may constrain their future supply. To overcome these potential constraints, cleaner production practices and more stringent environmental regulations must be adopted. The production of lithium is highly concentrated in Chile and Argentina, while vanadium production is mainly located in Russia and China. Consequently, special attention should be paid to the trade risk of those two metals with regard to the spatial mismatch of production and consumption. There is currently radical innovation taking place to develop new types of battery and other types of energy technologies (hydrogen cars) to replace electric vehicles. The use of most of those metals (excluding nickel) will come with huge resource losses in mining and at the end-of-life stage, and the in-use stock of cobalt, nickel, and vanadium is not largely attributed to any applications other than battery.

(2) **Photovoltaic (PV)-related metals**. PV is highly material intensive and relies on various critical metals. CdTe panels require cadmium and tellurium, while

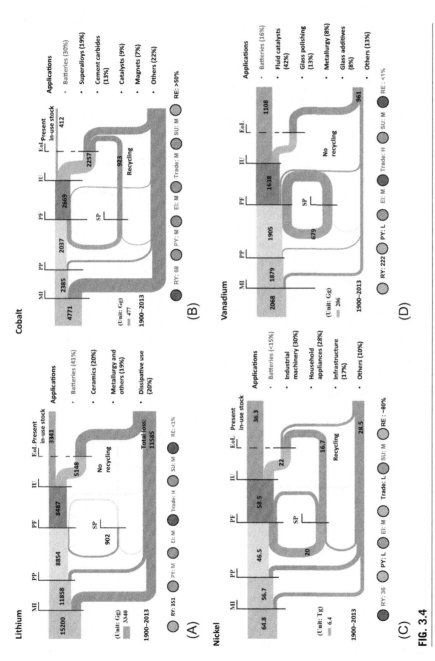

FIG. 3.4

Global material cycle of battery-related critical metals and key indicators of concern. Note: *EI*, environmental impact; *EoL*, end-of-life; *IU*, In-use; *MI*, mining; *PF*, product fabrication (manufacturing); *PP*, primary production; *PT*, production yield rate; *RE*, recycling rate; *RY*, reserve-to-production year; *SU*, substitutability.

Credit: Compiled by the author using data from Graedel, T. E., Harper, E. M., Nassar, N. T., Nuss, P., Reck, B. K. (2015). Criticality of metals and metalloids. Proceedings of the National Academy of Sciences, 112, 4257–4262.; Wang, P., Chen, L.-Y., Ge, J.-P., Cai, W., Chen, W.-Q. (2019). Incorporating critical material cycles into metal-energy nexus of China's 2050 renewable transition. Applied Energy, 253, 113612.; Zepf, V., Simmons, J., Reller, A., Ashfield, M., Rennie, C. (2014). Materials critical to the energy industry. An introduction. 2nd ed. BP Publisher, London, United Kingdom; and Philip, N., Eckelman, M. J. (2014). Life cycle assessment of metals: a scientific synthesis. PLos One, 9, (7), e101298.

a-Si panels are made using germanium, and CIGS panels need gallium, indium, and selenium. Notably, the major application of cadmium is the rechargeable nickel-cadmium batteries (80%), but this is selected here for its critical use in CdTe PV. The historical global cycle of those selected materials and their indicators of concern are illustrated in Fig. 3.5. The most visible pattern for most PV-related metals is the huge losses at the production stage. For each ton of material produced, around 71 tons of gallium, 152 tons of germanium, 1.86 tons of indium, 21 tons of tellurium, and 21 tons of selenium will be lost—mainly because they are by-products of their host metals' production. More efforts should be paid to high-yield rate technologies and practices in the primary production of these metals. The environmental impacts of the production of gallium, cadmium, and selenium are severe in terms of cumulative energy use, global warming potential, human health implications, and ecosystem damage. The production of germanium is highly concentrated in China, as it is a by-product of lead-zinc refining and coal ash production. China also controls most of the world's refined indium production. Almost all metals have a very poor circularity performance. Given that those metals are harder to refine and trade, metal recycling becomes a very promising approach for mitigating the supply risks. Substitution performance is modest for all metals given the rapid development in PV systems.

(3) **Wind or motor-related metals**. Wind turbine and motor production is heavily dependent on the use of rare earths, specifically neodymium, dysprosium, and praseodymium in powerful NdFeB permanent magnets, or samarium in SmCo magnets. Notably, the major application of samarium is the battery alloys (73%), but this is selected here for its potential use in wind turbines. The historical global cycles of those selected materials and their indicators of concern are illustrated in Fig. 3.6. In general, rare-earth minerals are not actually rare, and their reserves are abundant enough to meet future demand. The concerns about these rare-earth minerals are twofold: first, processing and separating the elements is technically difficult, environmentally hazardous, and consumes large amounts of energy. Second, the production of these minerals, especially the refining process, is now dominated by China, which has witnessed frequent disruptions in the past decades. At present, there is no record of rare-earth recycling activities worldwide, but some pilot projects have been launched in Europe. Notably, permanent magnets are also widely demanded for the production of electric motors, due to the expansion of electric vehicles. Still, presently, there are no suitable substitutes available for either neodymium or dysprosium in magnet production.

(4) **Vehicle-related (automotive catalyst) metals**. The electric automotive sector relies on a wide range of critical metals, including cobalt, lithium, nickel, palladium, platinum, and rare-earth elements (REEs). The autocatalyst sector is the biggest single consumer of platinum group metals known as PGMs—platinum, palladium, and rhodium used in various combinations, as well as lanthanum. The historical global cycles of these selected materials and their indicators of concern are illustrated in Fig. 3.7. There are huge reserves of all

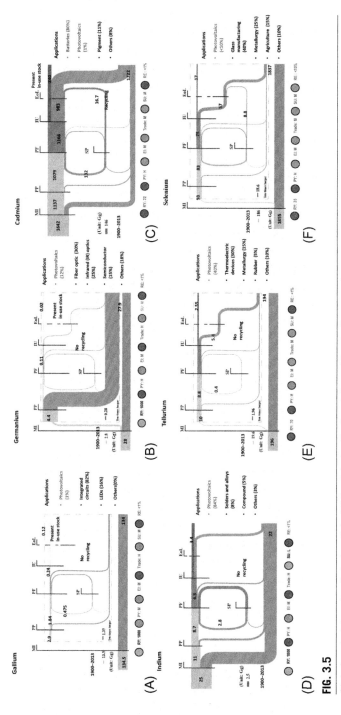

FIG. 3.5

Global material cycle of photovoltaic (PV)-related critical metals and key indicators of concern.

Credit: Compiled by the author using data from Graedel, T. E., Harper, E. M., Nassar, N.T., Nuss, P., Reck, B. K. (2015). Criticality of metals and metalloids. Proceedings of the National Academy of Sciences, 112, 4257–4262.; Wang, P., Chen, L.-Y., Ge, J.-P., Cai, W., Chen, W.-Q. (2019). Incorporating critical material cycles into metal-energy nexus of China's 2050 renewable transition. Applied Energy, 253, 113612.; Zepf, V., Simmons, J., Reller, A., Ashfield, M., Rennie, C., 2014. Materials critical to the energy industry. An introduction. 2nd ed. London, United Kingdom: BP Publisher; Allwood, J. M., Ashby, M. F., Gutowski, T. G., et al. (2011). Material efficiency: A white paper. Resources Conservation and Recycling, 55, (3), 362–381.; Philip, N., Eckelman, M. J. (2014). Life cycle assessment of metals: a scientific synthesis. PLos One, 9, (7), e101298; and USGS (2019). Mineral commodity summaries 2019. Reston, Virginia, United States: United States Geological Survey.

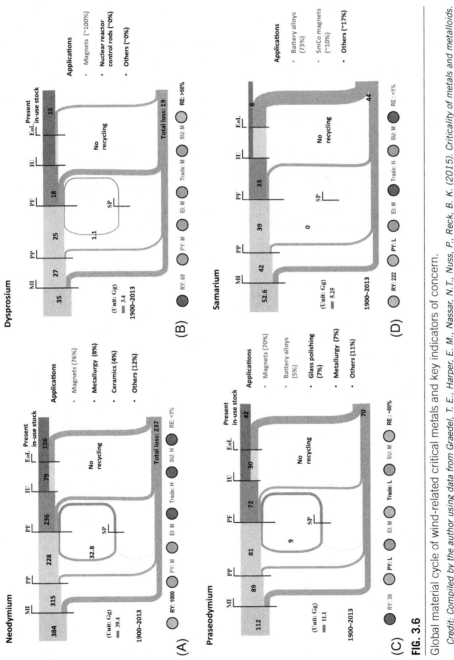

FIG. 3.6

Global material cycle of wind-related critical metals and key indicators of concern.

Credit: Compiled by the author using data from Graedel, T. E., Harper, E. M., Nassar, N.T., Nuss, P., Reck, B. K. (2015). Criticality of metals and metalloids. Proceedings of the National Academy of Sciences, 112, 4257–4262., Wang, P., Chen, L.-Y., Ge, J.-P., Cai, W., Chen, W. -Q. (2019). Incorporating critical material cycles into metal-energy nexus of China's 2050 renewable transition. Applied Energy, 253, 113612., and Zepf, V., Simmons, J., Reller, A., Ashfield, M., Rennie, C., 2014. Materials critical to the energy industry. An introduction. 2nd ed. London, United Kingdom: BP Publisher.

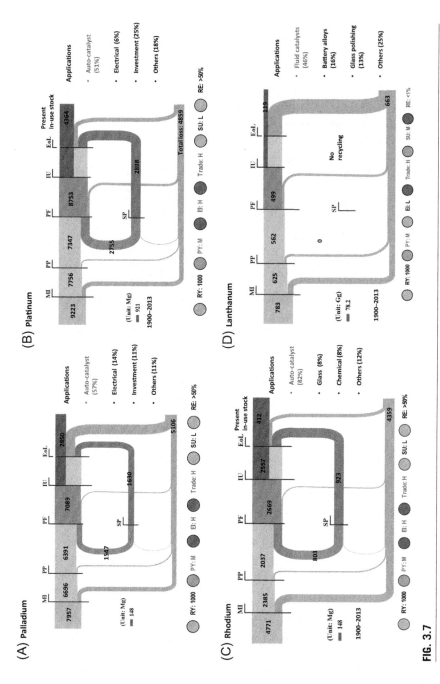

FIG. 3.7

Global material cycle of vehicle-related critical materials and key indicators of concern.

Credit: Compiled by the author using data from Graedel, T. E., Harper, E. M., Nassar, N. T., Nuss, P., Reck, B. K. (2015). Criticality of metals and metalloids. Proceedings of the National Academy of Sciences, 112, 4257–4262, Wang, P., Chen, L.-Y., Ge, J.-P., Cai, W., Chen, W.-Q. (2019). Incorporating critical material cycles into metal-energy nexus of China's 2050 renewable transition. Applied Energy, 253, 113612., and Zepf, V., Simmons, J., Reller, A., Ashfield, M., Rennie, C. (2014). Materials critical to the energy industry. An introduction. 2nd ed. London, United Kingdom: BP Publisher.

these minerals to meet future demand. However, international trade will be a major concern for PGMs as they are dominated by Russia and South Africa, which account for 86% of the total global production of platinum, 79% of palladium, and 92% of rhodium. The REE lanthanum is mainly supplied by China. The production of PGMs has a severe environmental impact and platinum compounds are highly bioactive. Among all these metals, rhodium causes the highest environmental burden per unit in terms of global warming potential, terrestrial acidification, and human toxicity (Philip & Eckelman, 2014). Autocatalysis technologies are also undergoing rapid innovation and are shifting from platinum to all PGMs, which can be substituted by other materials. With the expansion of electric vehicles, the corresponding technological innovation will make the automotive industry less reliant on PGMs, so that future demand for those metals is likely to decrease. The recycling performance of autocatalysts is known to be good, with rates of around 50%.

Global challenges call for cooperative action
Challenges: Pressing constraints beyond physical scarcity

(1) **Physical scarcity will not be a major concern**. Critical minerals are a finite, nonrenewable resource found in the Earth's crust. The long-term geological availability of mineral commodities is a key factor for determining potential resource depletion, so many institutions have conducted thorough investigations to estimate the reserve and the reserve base of different resources (Ober, 2017). Some consider depletion to be unavoidable given the fixed resource stock and continuous demand (Cohen, 2007; Gordon, Bertram, & Graedel, 2007; Ragnarsdóttir, 2008). For some energy-related minerals such as cobalt, nickel, and cadmium, geological scarcity might be a major concern. However, new technologies and innovation can offset the cost-increasing effects of depletion (Tilton & Lagos, 2007; Turner, 2008), and physical scarcity will not be a major concern for most metals.

(2) **Mining operation time lags and limited expansion rates**. Fast-growing demand requires a similar increase in mining capacity if there is no improvement in the efficiency of resource use. However, critical metals are produced through complex mining and refining processes, and the expansion of mining capacity faces two major challenges: (a) opening a new mine (and significantly scaling production) takes on average 10–20 years (Prior, Giurco, Mudd, Mason, & Behrisch, 2012) and (b) some metals are mined as coproducts or by-products of other host metals (Gunn, Graedel, & Espinoza, 2014). In this context, the recent studies (Bustamante & Gaustad, 2014; Kavlak, McNerney, Jaffe, & Trancik, 2015; Nassar, Wilburn, & Goonan, 2016) have pointed out that the limited scalability of mining capacity would pose a significant supply risk in the face of growing demand. Moreover, most of the metals in

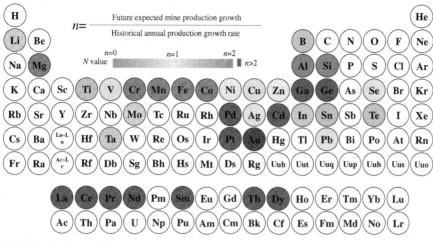

FIG. 3.8

Comparison of future and historical scalability of critical minerals.

Credit: Compiled by the author using data from Rietveld, E., Boonman, H., van Harmelen, T., Hauck, M., Bastein, T. (2019). Global energy transition and metal demand—An introduction and circular economy perspectives. The Hague, Netherlands: TNO, Netherlands Organisation for Applied Scientific Research.

this situation are critical metals in emerging applications, whose demand is expected to rise significantly, but the associated mining capacities are hard to expand at the same speed. Fig. 3.8 shows a comparison ratio between the future expected production growth rate (from 2011 to 2050) and the historical production growth rate of critical minerals. Demand for most energy-related minerals will rise faster than the historical trends, including lithium, cadmium, gallium, neodymium, dysprosium, tellurium, vanadium, and platinum. Hence, the future resource constraints facing those metals will be more severe.

(3) By-production constraints. Unlike major metals (e.g., iron, copper, and aluminum), critical metals do not often form their own deposits and are typically mined as the by-products (companion metals) of major metals (host metals). As defined in Nassar, Graedel, and Harper (2015), the companionability of critical metals can be quantifiably estimated by working out what percentage of their global primary production is the result of by-production Notably, the production of a companion metal is often unlikely to quickly respond to rapid changes in demand, and the production of companion metals is strongly dependent on their host metals, resulting in a lack of price elasticity. Moreover, many companion metals, like rare-earth elements, are located in a limited number of geopolitically concentrated ores. This complex linkage between by-products and host metals will, therefore, greatly constrain the future supply of most energy-critical minerals. In this context, improvements in mining and refining technologies can be very promising and may increase the recovery rate of host metals, which will further increase the supply of companion metals.

(4) **Mineral trade and geopolitics**. For centuries, concern about national energy security has mainly focused on fuels. However, with the energy system shifting from fuels to metals, clearly, national energy security will become closely linked to the geopolitics of minerals and metals. Unlike the fuels that are highly concentrated in a few nations/regions (e.g., Middle East, Australia, and the United States), the situation regarding critical minerals that are linked to energy technologies is highly complicated, and each mineral has its unique story (Eggert, 2011). For instance, global wind technologies rely on rare earths from China, but China also depends on cobalt from the Democratic Republic of the Congo for the electrification of its transport sector. Meanwhile, this dynamic new geopolitical landscape is affected by any changes in the exploration of new resources, technical innovation, recycling, environmental regulations, etc. Notably, the trade network of minerals, products, and waste will compound such challenges, especially given the current rise of trade protectionism. There is an ongoing tendency of political instability and the formation of monopolies and oligopolies in the critical mineral market. Thus, in this new era, national energy security strategies must pay special attention to the trade networks and geopolitics of critical metals.

(5) **Environmental impacts and social conflicts**. The mining and processing of critical minerals has serious environmental impacts: it generates toxic waste and pollutants, such as radioactive waste in REE processing that damage ecosystems and are transferred to other life-supporting systems (e.g., water and food). At present, critical minerals are mainly sourced from the developing and low-income nations in Asia, Latin America, and Africa, which have less technology available, low environmental awareness, and lax regulations. Cases of environmental damage caused by mineral mining and production are abundant, such as the artisanal cobalt mining in DRC (Celestin et al., 2018; Prause within this book), the radioactive waste generated by the rare-earth plant in Malaysia (Law, 2019), and illegal REE mining and processing in China (Lee & Wen, 2018). Although Graedel et al. (2012) consider environmental impact to be a key indicator for determining material criticality, there is a lack of in-depth investigation (and localized data) on the environmental impact of mineral processing, and the severity of environmental impacts on the supply of critical minerals. Clearly, renewable energy can help the world move toward a cleaner and more sustainable future, but at the cost of polluting those mineral processing areas, which are usually poor and remote places. Moreover, some energy-critical minerals are listed as 'conflict minerals,' which are extracted in conflict zones and sold to perpetuate the fighting. The risks caused by environmental impacts and social conflicts, albeit not widely noticed, can further constrain the future supply of those energy-critical minerals.

Looking ahead: Joint efforts by stakeholders throughout the metal cycle and trade network

How can we mitigate such pressing metal constraints on the global energy system transition? From a systematic perspective, we clarified this challenge and its corresponding metal-related strategies in Fig. 3.9. As shown in Fig. 3.9A, if a business-as-usual strategy is followed, the future energy transition will cause unprecedented demand for metals, which will increase the environmental impacts and various conflicts associated with metal production and trade. Clearly, if fewer minerals are extracted, there will be fewer conflicts and environmental damage. Hence, there is a need for joint efforts from all stakeholders throughout the material cycle to expand the metal supply and improve metal efficiency, so that less metal is required for same technical functions. These efforts can be divided into three potential sets of strategies as shown in Fig. 3.9B.

   (1) More sustainable metal supply. Clearly, more primary metal resources will be required to meet unprecedented future needs. Thus, there is a need for stakeholders, mainly at the mining and production stage, to (a) expand geological exploration to increase the resource base and (b) improve their mining and processing technologies to be able to harvest more metal resources. However, it's worth pointing out that this set of strategies may cause the kind of severe environmental impacts that are associated with metal extraction and production. Stakeholders—especially local governments and industries—should pay careful attention to impact mitigation and environmental protection. Most energy-critical metals are applied to clean and green technologies at the consumer side, while pollution is left at the production side. To overcome the resulting environmental inequality, cooperative mechanisms are needed to actively share the burden between both the production and the consumption sides.

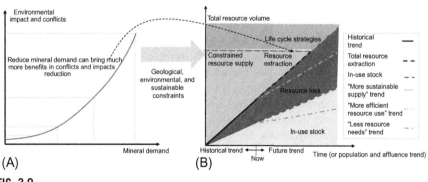

(A)   **(B)**

FIG. 3.9

Strategies to mitigate critical metal constraints on the future energy system transition. (A) Relationship of conflicts and mineral demand, (b) strategies for mineral use under constraints.

Credit: Compiled by the author.

(2) More efficient metal system. The material system plays a critical role in linking the metal supply to energy applications. Through our material flow analysis, we found that tremendous resource losses exist at each stage in the material cycle, especially at the end-of-life stage. To reduce resource losses, strategies can be implemented at each stage in the material cycle, e.g., cleaner production at the production stage, material efficiency at the manufacturing stage, and higher material circularity at the end-of-life stage. However, such scattered strategies become problematic, even contradictory, in relation to critical metals. Given that critical metals are widely used, yet applied in extremely small amounts, recycling may not be economically or environmentally feasible (see Weil et al. within this book). Sometimes, if products are designed to be recycled, it can make them less durable and cause more waste flows. Hence, there is a need to incorporate all stakeholders throughout the material cycle to systematically improve processes and set priorities.

(3) Fewer material needs. Strategies for material reduction focus on technical innovations that reduce the amount of material required by energy technologies during the in-use stage. At the energy infrastructure level, better management of infrastructure can help to deliver more renewables services and prolong the lifetime and efficiency of renewable energy technologies during the in-use stage. Consequently, less new infrastructure would be required, and critical materials can be saved. At the material level, more advanced technologies can help to reduce the amount of critical metals required, which means more research and development efforts should be directed toward improving those technologies.

In addition to strategies based on the material cycle, there is also an urgent need to promote a well-functioning trade network to underpin unimpeded flows of metal-contained minerals, products, and technology across national boundaries (Eggert, 2011). Thus, the key to the global energy transition is cooperation between nations that produce renewable energy technologies and nations endowed with an abundance of minerals. The present emphasis of global energy governance is mainly about the supply risks involved with oil, coal, and other fossil producers, while renewable technologies were widely considered to enhance global energy security [e.g., in the IEA report (Ölz, Sims, & Kirchner, 2007)]. However, critical metals are not equally distributed across national borders, but are highly concentrated and controlled by a limited number of nations or producers (EU Commission, 2017), which could pose a threat to the global renewable transition. Consequently, thinking about the nexus between metals and energy can help to identify the potential supply risks and enhance international energy cooperation and policy making. Meanwhile, nations should make joint interventions to reduce trade barriers and thereby promote a successful energy transition for a more sustainable future.

Acknowledgments

We appreciate the efforts of Ruru Han and Shen Zhao, who helped us prepare the materials for this work. The chapter is based on our presentation titled "Material Cycle and its Energy,

Environment, and Geopolitics Nexus" from an internal seminar at the IUE, and we thank all the participants for their comments that have enriched this work. This chapter is supported by the National Natural Science Foundation of China (no. 41671523, no. 71904182).

References

American Physical Society. (2011). *Energy critical elements: Securing materials for emerging technologies.* American Physical Society and Materials Research Society joint report. D.C.: Washington.

Bustamante, M. L., & Gaustad, G. (2014). Challenges in assessment of clean energy supply-chains based on byproduct minerals: A case study of tellurium use in thin film photovoltaics. *Applied Energy, 123*, 397–414.

Celestin, B. L. N., Lidia, C., Vincent, H., Thierry, D. P., Nelly, D. S., & Tony, K. (2018). Sustainability of artisanal mining of cobalt in DR Congo. *Nature Sustainability, 1*(9), 495.

Chen, W. Q., & Graedel, T. E. (2012). Anthropogenic cycles of the elements: A critical review. *Environmental Science and Technology, 46*, 8574–8586.

Cohen, D. (2007). Earth's natural wealth: an audit. *New Scientist*, 34–41.

Davis, S. J., Lewis, N. S., Shaner, M., Aggarwal, S., Arent, D., Azevedo, I. L., et al. (2018). Net-zero emissions energy systems. *Science, 360*(6396), 1–9.

Du, X., & Graedel, T. E. (2011). Uncovering the global life cycles of the rare earth elements. *Scientific Reports, 1*, 1–4.

Duclos, B. S. J., Otto, J. P., & Konitzer, D. G. (2009). Design in an era of constrained resources. *Mechanical Engineering, 132*, 36–40.

Eggert, R. G. (2011). Minerals go critical. *Nature Chemistry, 3*(9), 688–691.

EU Commission. (2010). *Critical raw materials for the EU 2010.* Brussels: EU Commission.

EU Commission. (2014). *Critical raw materials for the EU 2014.* Brussels: EU Commission.

EU Commission. (2017). *2017 list of critical raw materials for the EU.* Brussels: EU Commission.

Frenzel, M., Kullik, J., Reuter, M. A., & Gutzmer, J. (2017). Raw material "criticality"—Sense or nonsense? *Journal of Physics D: Applied Physics, 50*(12), 123002.

Gordon, R. B., Bertram, M., & Graedel, T. E. (2007). On the sustainability of metal supplies: A response to Tilton and Lagos. *Resources Policy, 32*, 24–28.

Graedel, T. E., Barr, R., Chandler, C., Chase, T., Choi, J., Christoffersen, L., et al. (2012). Methodology of metal criticality determination. *Environmental Science and Technology, 46*, 1063–1070.

Graedel, T. E., Harper, E. M., Nassar, N. T., Nuss, P., & Reck, B. K. (2015). Criticality of metals and metalloids. *Proceedings of the National Academy of Sciences, 112*, 4257–4262.

Graedel, T. E., & Reck, B. K. (2016). Six years of criticality assessments: What have we learned so far? *Journal of Industrial Ecology, 20*, 692–699.

Gunn, G., Graedel, T. E., & Espinoza, L. T. (2014). Metal resources, use and criticality. In *Critical metals handbook.* Hoboken, New Jersey: Wiley Pubisher.

Habib, K., & Wenzel, H. (2016). Reviewing resource criticality assessment from a dynamic and technology specific perspective—Using the case of direct-drive wind turbines. *Journal of Cleaner Production, 112*, 3852–3863.

Hatayama, H., & Tahara, K. (2015). Evaluating the sufficiency of Japan's mineral resource entitlements for supply risk mitigation. *Resources Policy, 44*, 72–80.

Hayes, S. M., & Mccullough, E. A. (2018). Critical minerals: A review of elemental trends in comprehensive criticality studies. *Resources Policy, 59*, 192–199.

IEA. (2018). *World Energy Outlook 2018*. Paris, France: International Energy Agency.

IEA. (2019). *Global EV Outlook 2019: Scaling up the transition to electric mobility*. Paris, France: International Energy Agency.

Ioannidou, D., Heeren, N., Sonnemann, G., & Habert, G. (2019). The future in and of criticality assessments. *Journal of Industrial Ecology, 4*(23), 751–766.

IRENA. (2017). *Renewable capacity statistics 2017*. Abu Dhabi: International Renewable Energy Agency.

IRENA. (2019). *Global energy transformation: A roadmap to 2050*. Abu Dhabi: International Renewable Energy Agency.

Jin, Y., Kim, J., & Guillaume, B. (2016). Reviewing critical material studies. *Resources, Conservation and Recycling, 113*, 77–87.

KAPSARC. (2018). *China's policy drivers of future energy demand*. Riyadh, Saudi Arabia: King Abdullah Petroleum Studies and Research Center.

Kavlak, G., & Graedel, T. E. (2013a). Global anthropogenic tellurium cycles for 1940-2010. *Resources, Conservation and Recycling, 76*, 21–26.

Kavlak, G., & Graedel, T. E. (2013b). Global anthropogenic selenium cycles for 1940–2010. *Resources, Conservation and Recycling, 73*, 17–22.

Kavlak, G., McNerney, J., Jaffe, R. L., & Trancik, J. E. (2015). Metal production requirements for rapid photovoltaics deployment. *Energy & Environmental Science, 8*, 1651–1659.

Law, Y.-H. (2019). Politics could upend global trade in rare earth elements. *Science, 364*, 114–115.

Lee, J. C. K., & Wen, Z. (2018). Pathways for greening the supply of rare earth elements in China. *Nature Sustainability, 1*, 598–605.

Licht, C., Peiró, L. T., & Villalba, G. (2015). Global substance flow analysis of gallium, germanium, and indium: Quantification of extraction, uses, and dissipative losses within their anthropogenic cycles. *Journal of Industrial Ecology, 19*, 890–903.

Løvik, A. N., Hagelüken, C., & Wäger, P. (2018). Improving supply security of critical metals: Current developments and research in the EU. *Sustainable Materials and Technologies, 15*, 9–18.

Løvik, A. N., Restrepo, E., & Müller, D. B. (2015). The global anthropogenic gallium system: Determinants of demand, supply and efficiency improvements. *Environmental Science and Technology, 49*, 5704–5712.

Nassar, N. T., Graedel, T. E., & Harper, E. M. (2015). By-product metals are technologically essential but have problematic supply. *Science Advances, 1*(3), 1–11.

Nassar, N. T., Wilburn, D. R., & Goonan, T. G. (2016). Byproduct metal requirements for U.S. wind and solar photovoltaic electricity generation up to the year 2040 under various Clean Power Plan scenarios. *Applied Energy, 183*, 1209–1226.

NRC (National Research Council). (2008). *Minerals, critical minerals, and the U.S. economy*. Washington, D.C.: National Academies Press.

Ober, J. A. (2017). *Mineral commodity summaries 2017*. Reston, Virginia, United States: US Geological Survey.

Ölz, S., Sims, R., & Kirchner, N. (2007). *Contribution of renewables to energy security*. Paris, France: International Energy Agency.

Philip, N., & Eckelman, M. J. (2014). Life cycle assessment of metals: a scientific synthesis. *PLos One, 9*(7), e101298.

Prior, T., Giurco, D., Mudd, G., Mason, L., & Behrisch, J. (2012). Resource depletion, peak minerals and the implications for sustainable resource management. *Global Environmental Change, 22*, 577–587.

Ragnarsdóttir, K. V. (2008). Rare metals getting rarer. *Nature Geoscience, 1*, 720–721.

Rietveld, E., Boonman, H., van Harmelen, T., Hauck, M., & Bastein, T. (2019). *Global energy transition and metal demand—An introduction and circular economy perspectives.* The Hague, Netherlands: TNO, Netherlands Organisation for Applied Scientific Research.

Rosenau-Tornow, D., Buchholz, P., Riemann, A., & Wagner, M. (2009). Assessing the long-term supply risks for mineral raw materials—a combined evaluation of past and future trends. *Resources Policy, 34*, 161–175.

Skirrow, R. G., Huston, D. L., Mernagh, T. P., Thorne, J. P., Duffer, H., & Senior, A. (2013). *Critical commodities for a high-tech world: Australia's potential to supply global demand.* Canberra: Geosicence Australia.

Smil, V. (2010). *Energy transitions: History, requirements, prospects.* Santa Barbara: Praeger Publisher.

Tilton, J. E., & Lagos, G. (2007). Assessing the long-run availability of copper. *Resources Policy, 32*, 19–23.

Turner, G. M. (2008). A comparison of the limits to growth with 30 years of reality. *Global Environmental Change, 18*, 397–411.

U.S. DoE. (2010). *Critical materials strategy 2010.* Washington DC: United States Department of Energy.

U.S. DoE. (2011). *Critical materials strategy 2011.* Washington DC: United States Department of Energy.

UNEP. (2009). *Critical metals for future sustainable technologies and their recycling potential.* Nairobi, Kenya: United Nations Environment Programme.

Wang, P. (2019). *Dynamic analysis of anthropogenic metal cycle: Resource efficiency and potential strategies.* PhD Thesis University of New South Wales.

Wang, P., Chen, L.-Y., Ge, J.-P., Cai, W., & Chen, W.-Q. (2019). Incorporating critical material cycles into metal-energy nexus of China's 2050 renewable transition. *Applied Energy, 253*, 113612.

Wang, P., & Kara, S. (2019). Material criticality and circular economy: necessity of manufacturing-oriented strategies. *Procedia CIRP, 80*, 667–672.

WWF. (2014). *Critical materials for the transition to a 100% sustainable energy future.* Gland, Switzerland: World Wide Fund for Nature.

Zepf, V., Simmons, J., Reller, A., Ashfield, M., & Rennie, C. (2014). *Materials critical to the energy industry. An introduction* (2nd ed.). London, United Kingdom: BP Publisher.

The dependency of renewable energy technologies on critical resources

4

Volker Zepf

Independent consultant for Geography and Resource Strategies, Augsburg, Germany

Introduction to the "critical," "resources," and "renewables"

There is a difference between the terms "critical resources," "critical metals," and "critical materials" even though they are often used synonymously. Rather, they are subsets of each other: a critical metal is a critical material, which is a critical resource. "Criticality" in general is seen as a measure of the economic importance of a raw material and the associated potential supply risk. Today, criticality primarily addresses the material demand involved in the production of advanced technologies, energy technologies, and green or clean technologies (see also Wang et al. and Gilbert within this book).

The need for renewable energy technologies has become apparent and is continuously gaining momentum, particularly due to increasing awareness about climate change and the potential options mankind has for contributing to a smooth transition to the future. Fears emerged because numerous specialized materials are required to achieve the desired functionality and efficiency of these advanced technologies. The unavailability of these materials could thus hamper the desired clean, future development.

The fears are partially valid, as several of these technologies indeed require a huge variety of elements and materials, but at the same time, several false inductive conclusions led to incorrect understandings, estimates, and projections about demand. A short example, which is discussed in more detail in the following paragraphs, are the figures shown in a Climate Smart Mining Infographic, where a 3 MW rated wind turbine requires 2 tons of rare earth elements (World Bank, 2019). This is only true for one type of wind turbine, which accounts for about 20% of the total installed wind power (Zepf, 2012). A further inaccuracy emerges, as the infographic refers to rare earth elements (REE), a group of 17 individual elements on the periodic table as if there was only one element; whereas the author probably only had the magnetic materials neodymium, praseodymium, and maybe dysprosium in mind. REEs vary greatly in terms of availability and criticality, so this inaccuracy spreads confusion.

The Material Basis of Energy Transitions. https://doi.org/10.1016/B978-0-12-819534-5.00004-0

REEs will be used several times as an example in this chapter, as a lot of problems associated with criticality and the resource nexus can be shown through the REEs.

This short example demonstrates the complexity of criticality; it cannot be reduced to a few generic numbers and a mere quest for the "best system."

This chapter analyzes renewable energy systems in relation to their need for raw materials. The systems discussed here were chosen according to the classifications published by the International Energy Agency (IEA) (see next section).

Systemic and considerable size differences exist between the different energy systems. Therefore, as shown in the above example, purely quantitative analysis of materials demand is not possible and not useful. Instead, a qualitative narrative will help to explain the dedicated materials demand.

Thus, in this chapter, the "*Stoffgeschichten*" (Story of Stuff) approach suggested by Reller (Böschen, Soentgen, & Reller, 2004) will be applied for the discussion of the various energy systems. This is a narrative and qualitative method, which is accompanied by quantitative data to illustrate the narrative. An element or system will be analyzed along its life cycle starting from extraction, through to separation, refining, production, and use. The analysis also includes the re-phases and the end-of-life phases. This chain will be looked at in relation to various aspects, primarily economic, ecological, social, and political. The future outlook is chiefly based on the research of the author (Zepf, Reller, Rennie, Ashfield, & Simmons, 2014).

The next sections continue with a discussion of the terms "renewable energy technologies," "criticality," and "resources." In "Renewable energies and their material resources demand" section, the selected energy technologies and their resource requirements will be analyzed and discussed in detail. Examples will demonstrate how the quantity of materials differs between the technologies. The conclusion summarizes aspects that should be considered when dealing with different renewable energy technologies and their specific materials demands.

Renewable energy technologies

"Renewable energy technologies" is an umbrella term that stands for energy production using a renewable energy source like solar, wind, water (hydro and tidal), biomass (biofuels and wastes), and geothermal heat. The International Energy Agency (IEA) calls these "renewable energy sources" (IEA, 2019: 3). A "renewable energy system" is then seen as the actual power plant that converts the renewable energy carrier or source into electrical, mechanical or thermal energy for use by the consumer. In 2017 renewable energy accounted for 13.6% of the global energy market, of which biofuels and wastes contributed 67.9%, followed by hydro (18.5%), wind (5.1%), geothermal (4.5%), and solar and tidal combined (3.9%) (IEA, 2019: 3).

Biofuels and waste, referred to here as biomass, will only be briefly addressed as they are not deemed to entail any demand for critical materials. However, unlike all other renewable energy systems, biomass requires continuous material input in order to operate, so that its side effects may determine criticality here. Wind energy is usually differentiated between onshore and offshore technologies, however, with

Renewable energy technologies*					
Biomass	Wind	Water hydro	Water tidal	Solar	Geothermal
Biomass thermal	Onshore wind	Run-of-river	Tidal wave	Photovoltaics (PV)	Geothermal electric
Biomass fermen-tation	Offshore wind	(Pumped) storage	Sea current	Solar thermal	Geothermal heat
			Wave vertical movement	Concentrated solar power (CSP)	

* Renewable Energy Technologies covered in this chapter.
Selection based on International Energy Agency. Renewables information. Overview 2019.

FIG. 4.1

Renewable energy technologies.

Credit: Own source based on International Energy Agency, Renewables Information, October 2019.

regard to material issues, it is more useful to differentiate between direct-drive permanent magnet technologies and all others (for details see "Renewable energies and their material resources demand" section). Water energy is divided into hydro and tidal, whereby hydro is further differentiated between run-of-river and storage power plants, and tidal energy systems are distinguished between tidal wave, sea current, and wave power systems. Solar energy is subdivided into photovoltaics, solar thermal, and concentrated solar power systems. Lastly, geothermal power plants provide heat for heating and/or the production of electricity. Fig. 4.1 provides a basic overview of the systems covered in this chapter.

The "criticality" aspect

When the attribute "critical" is applied to a certain element or raw material, it is primarily seen as a function of supply risk (SR) and economic importance (EI). In 2008 the US National Research Council (NRC) published a study about critical minerals and the US economy. The methodology used to determine criticality was based on two key aspects: (1) the impact that a supply restriction would have on the economy, and (2) the general threat of supply risk. Both aspects were divided into several categories, such as substitutability or monopolistic market situations (NRC, 2008). This study has been succeeded by several further studies in the United States and Europe that feature variations of the basic idea of criticality (European Commission, 2018; Fortier et al., 2018; US DoE, 2010), but have different foci, such as the US economy (NRC, 2008), the European Economy or the energy industry (Zepf et al., 2014). The frequent revisions of criticality lists show the evolution and the changing importance of technologies, and with them, the changing materials demands and notions of criticality. The European Critical Raw Material Initiative (EU CRM) published the first criticality list in 2011. A total of 41 elements selected by experts were investigated and 14 of those were identified as critical for the European Economy. The first revision in 2014 looked at 54 elements, of which 20 were listed as critical, then in 2017 a second revision analyzed 61 elements and the criticality list was expanded to include 26 elements (see also Wang et al. and Gilbert within this book).

These studies analyze materials based on expert input, whereas Zepf et al. (2014) provided another structured approach in a study about critical materials in the energy industry. A funnel system with several filters was used to determine criticality. This made it possible to start with the entire periodic table and then apply various filters that can be adapted to the respective economy to generate a criticality list. In that study, 23 elements were determined to be critical. Fig. 4.2 provides an overview of a selection of studies, including which elements were analyzed and which were identified as critical (see also Fig. 3.2 by Wang et al. within this book). The industrial and natural materials included in the EU CRM list have been excluded from the figure for ease of display.

Usually not covered in criticality lists is the factor of transnational companies, which partially control decisive shares of global raw materials production. Glencore, for example, is responsible for approximately 30% of the global production of copper, cobalt, lead, and zinc, and about 60% of global nickel production (Glencore, 2018). Raw material trade also serves as an ideal arena for financial products like shares, exchange trade funds, and possibly speculations. Many of these products may lay outside the originally envisaged supply risk aspects.

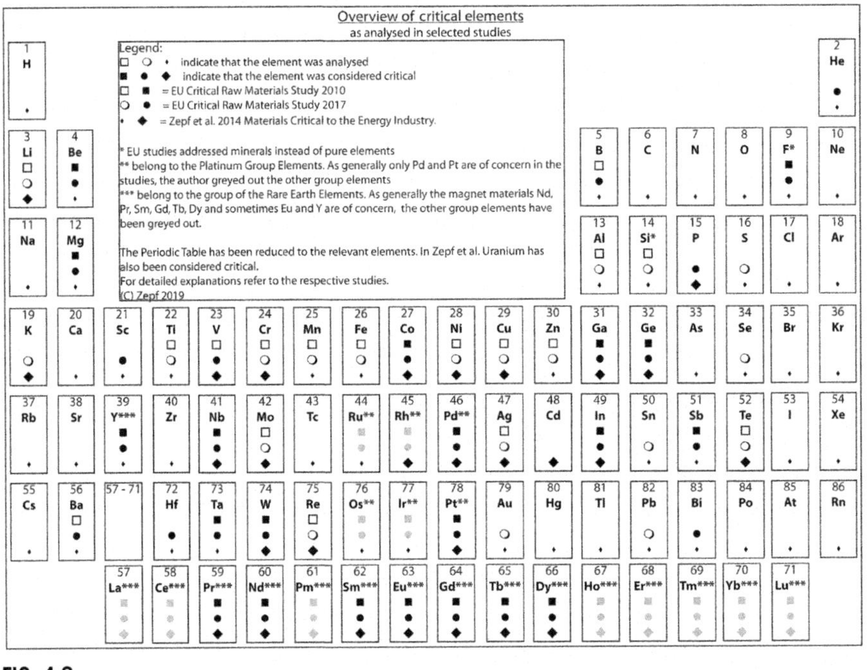

FIG. 4.2

Overview of critical elements.

Credit: Own Source, © Zepf 2019.

Recycling is usually also covered in the subset of supply risk; however, there is a difference between recyclability and actual recycling. UNEP provides an overview of end-of-life (EoL) recycling rates for the elements of the periodic table (UNEP, 2013: 48). Even though there are lots of research and knowledge about recycling methods, the cost of many raw materials is still too low to make recycling competitive. Adequate recycling plants do not exist and the required quantitative and qualitative scrap input is not guaranteed (Yang et al., 2016).

Ultimately, criticality is not an indicator of geological scarcity, but rather of unattainability. REEs, for example, are abundant and have reserves (120 million tons) to production (170,000 tons) ratio of approximately 700 years (USGS, 2019a). Thus the problem is not physical availability, but the will to invest in new mines and separation technology. This is costly and it takes several years to get all the permits, licenses, finances, and socioeconomic issues solved before that first cent can be earned. Criticality is, therefore, a very volatile issue and not at all a uniformly agreed-upon set of data.

Material and nonmaterial resources

According to the definition given by the US Geological Survey (USGS), "resources" are raw materials that are naturally occurring in or on the Earth's crust (USGS, 2019b). In this sense, materials can be of organic or nonorganic nature. The nonorganic materials can further be divided into mineral resources, such as ores, industrial rocks and minerals, and energy resources such as coal, oil, and gas. Organic materials can be seen as reproducible within a manageable timespan, whereas nonorganic materials require millions of years for their development and therefore cannot be considered renewable. But except for fossil fuels, all other mineral resources can at least theoretically be recycled and thus lack this issue of finiteness. Fossil fuels are irreversibly lost when they are used. The physical finiteness and availability of these materials should be of concern, yet, with the exception of uranium, fossil fuels do not appear on the criticality lists.

Global material extraction data provides a clue about the magnitude and what we can expect in the future. In 1950 approximately 15 billion tons of materials were extracted and in 2005 this figure was 60 billion tons. It is estimated that in 2050 about 140 billion tons will be extracted (Krausmann et al., 2009). Alongside the sheer quantity of extracted materials, another decisive material factor is the diversity of the elements produced. Achzet et al. (2011) showed the evolution of elements widely used in energy pathways. While in the 1700s just a few elements such as carbon, calcium, and iron were used for energy systems, today nearly the entire periodic table is in use thanks to the development of the steam engine and myriad other inventions. These additional elements provide some decisive advantages, as they facilitate new functionalities. For example, lithium can be used as a carrier substance in batteries and makes it possible to produce more powerful batteries, which are needed for electro mobility (see also Weil et al. within this book). Another advantage of using all these elements is the possibility of achieving a miniaturization of systems. REE-based permanent magnets make it possible to construct tiny vibration motors that weigh < 1 g and are used in today's sleek smartphones. Alloying steel with a low percentage of yttrium results in a firm but lightweight material that is used in the automotive industry. The resulting weight reduction reduces gasoline consumption and thus contributes to energy efficiency.

In addition to these material aspects, a second nonmaterial aspect should be kept in mind: knowledge, education, procedures, patents, time, etc. are indispensable prerequisites for the development, extraction, and manufacture of products. The REEs example provides proof of this, too. From 2005 to 2008 China published by far the most scientific papers about REE separation, recycling, the impact of REE mining and separation on the environment, and other REE-related topics (Adachi, Imanaka, & Tamura, 2010). It turned out that China's knowledge about the extraction and production of REEs was better than anywhere else in the world. When the need for REEs became apparent in around 2010, the knowledge deficit in the Western world led to decisive efforts in the United States, Japan, and Europe to build a knowledge base that would enable REEs to be produced outside China. Recently a further study revealed that China has another technological edge concerning patents around REEs: since 1950 China has filed 25,000 patents, whereas Japan has filed 14,000 followed by the United States with 10,000, and Europe with 7000 (numbers rounded) (Ng, 2019). This shows the increasing disparity in knowledge that may influence criticality, yet this factor is not considered in the criticality studies mentioned above.

Renewable energies and their material resources demand

All of the renewable energy systems have to be analyzed along with their life cycles so that the various materials and resource demands can be deducted at the respective points in time. Fig. 4.3 sketches the basic steps: the life cycle starts with system manufacturing and the construction of a power plant or installation of an energy system, proceeds to the time of operation, until eventually, the system is subject to disassembly at the end of the operative phase. Other phases such as exploration have been excluded because they are not considered to have a significant impact on resources. For all of the basic phases, there are various prevailing resource issues. "System manufacturing" highlights that the production of wind turbines and solar cells is the first step before they are installed at a site. The construction phase will result in the production of emissions (e.g., caused by the machinery), immissions, solid and liquid wastes, etc., which have to be treated accordingly to keep the entire system and its life cycle as environmentally friendly as possible.

The operative phase of renewable energy production obviously does not require a substantial additional feed of materials besides maintenance input, although there is one exemption: biomass. Eventually, the power plants will reach their EoL and may be disassembled and recycled. Here, a final resource nexus shows up as several of the power plants contain critical materials that are possibly rare and/or partially toxic, so that specialized recycling technologies have to be applied. This is especially true of

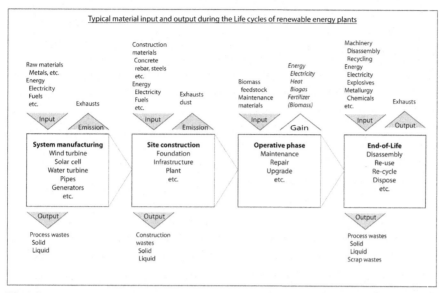

FIG. 4.3

Typical material input and output during life cycles of renewable energy plants.

Credit: Made by the author.

some types of solar cells that contain cadmium, tellurium, or arsenide. It should not be forgotten that the lifespans of these systems are usually considerable: wind turbines and solar cells are designed to operate for at least 20 years, whereas hydroelectric plants in dams operate for several decades. Thus recycling always has to consider the time factor in relation to the availability of input material for recycling. Fig. 4.3 does not include an analysis of the grid and nonmaterial resources.

Biomass

A decisive reason for mentioning biomass is that energy production from biomass is baseload capable. When gas is being produced, it can be stored and used at a later stage. This renders biomass a major pillar of renewable energy supply. The common types of biomass plants are *Biomass Fermentation Plants*, which produce electric energy or gas, and *Biomass Thermal Power Plants*. These plants are standard constructions with no major critical materials requirements both in terms of quality and in quantity.

However, biomass power plants need a continuous supply of biomass. These fuels have to be actively produced all year long. As not all the biomass is available throughout the year, so biomass plants need to include storage facilities or be designed to process a wide range of biomass fuels. A major factor is, therefore, the sustainability of the biomass production and whether this production is competing with food production. The production of food may be sacrificed in favor of biomass production, particularly when the latter is subsidized. If biomass growth involves monocultures and is promoted by the use of fertilizers, herbicides, pesticides, and fungicides, the overall clean balance of biomass energy may become quite unsustainable.

All the positive and negative effects are known and biomass energy systems are only suitable in areas where enough biomass can be produced. Areas with low vegetation, very short vegetation times and water stress cannot sustain biomass plants. So the future development of biomass plants is difficult to estimate. Given that no or only irrelevant quantities of critical elements and metals are required for the construction and operating phase of biomass plants, future constraints concerning construction materials and recycling are unlikely.

Water

Water energy used to produce electricity is generally created by damming and directing water through turbines that are connected to generators which produce electric energy. This principle is found in all hydroelectric power plants (PP), although there are some differentiations in design, size, and operating technique.

The major hydroelectric power plants are of the run-of-river and (pumped) storage type, which operate primarily in freshwater conditions. Water use in tidal systems is differentiated as "tidal wave," "sea current," and "wave vertical movement" power plants, which operate in saltwater conditions and are thus extremely prone to corrosion.

Tidal, run-of-river, and reservoir type PPs are baseload capable, provided that water levels are suitable and persistent. Low water levels due to droughts and floods caused by extreme precipitation events, which may be increasing in severity due to climate change, pose future challenges and potential shortfalls for some hydroelectric plants. Pumped storage type plants use surplus energy to pump water into dedicated storage reservoirs at higher elevations so that the potential energy can be recalled whenever needed.

Run-of-river and reservoir type hydroelectric power plants

Classic water energy is delivered by run-of-river and reservoir type hydroelectric PPs. These are usually located along rivers where the continuous flow of freshwater is fed through turbines that are housed in dams built across the river. The turbines drive generators in adjacent buildings to produce electricity, which is subsequently converted to meet grid requirements. In a few cases, the generator system is actually built into the dams or even mountains (Verbund, 2019). The dividing line between the two plant types is blurred. A dam can be built in such a way that the river morphology remains unchanged, or is only marginally changed, such as the run-of-river PPs along the Danube and alpine rivers. On the other hand, dam construction may involve retaining upstream water in a larger area, so that a lake will develop. Prominent examples for such reservoir type PPs are the Three Gorges Dam in China and the Glen Canyon Dam located in the United States. Smaller examples can be found in alpine regions, like the Austrian reservoirs Wasserfallboden and Mooserboden. These two are part of the large and complex Tauern Hydroelectric Power Plant System located near the village of Kaprun (Verbund, 2019).

Obviously the size and thus materials demands of the entire constructions and the PPs themselves vary in magnitude. The Three Gorges Dam is 2309 m long and 181 m high, which makes the construction one of the largest in the world. It used 27 million m^3 of concrete for the construction of the dam and the PP. It also incorporated 32 Francis turbines and associated main generators, each rated with 700 MW (Niu, 2016), that provide a total of 22.4 GW of installed power, equivalent to about 22 nuclear power plants. In contrast, the Kaprun Main Stage PP is rather modestly sized. The storage PP Limberg I features 2 Francis turbines and generators of 112 MW each. The adjacent Wasserfallboden dam is a double-curved arch-gravity dam with a crest length of 357 m, a width ranging between 6 and 37 m, and a dam height of 120 m. The construction required 446,000 m^3 of concrete (Verbund, 2019). Table 4.1 provides an overview of the materials data.

Next to the bulk concrete volumes and the large amounts of rebar that are needed for the construction, the turbines and generators are key parts and incorporate both sturdy low-end, but also high-tech materials, shapes, and designs. In principle, three different turbine types are being used nowadays: Francis turbines are the most

Table 4.1 Hydroelectric power plants—Selected technical and materials-related data.

Hydroelectric project	Three Gorges dam	Glen Canyon dam	Kaprun main stage Wasserfallboden	Danube border power plant Jochenstein
System type	Gravity dam	Concrete arch	Double-curved arch gravity dam	Run-of-river, slightly curved
Country	China	United States	Austria	Germany
Years of construction	1994–2012	1957–64	1952	1952–56
Number of main turbines with capacity	32 × 700 MW	8 × 165 MW	2 × 112 MW	5 × 28.9 MW
Total installed capacity	22,400 MW	1320 MW	280 MW	132 MW
Annual power generation	87,000 GWh	5000 GWh	500 GWh	850 GWh
Dam height	181 m	216 m	120 m	20 m
Dam length	2309 m	475 m	357 m	240 m
Concrete volume	27.2 million m^3	4.1 million m^3	0.5 million m^3	Unknown
Steel used	0.5 million tons	Unknown	Unknown	Unknown
Data source	Niu (2016)	USBR (2019)	Verbund (2019)	Verbund (2019)

commonly used water turbines and offer high-efficiency factors of about 90%. Their rated power ranges from 10 kW to 700 MW, so such turbines may have diameters of as much as 10 m, heights of 5 m, and weigh up to 150 tons. The unique advantage of the Francis turbine is its capability to be used as a pump as well, making it ideally suited for use in pumped storage applications. For lower water pressures, the Kaplan turbine is the most suitable choice. It resembles a ship propeller and can handle low pressures, as well as high flow rates and variable water flows. It reaches efficiencies of up to 96% and is often used in run-of-river power plants. A third type is the Pelton turbine, which consists of up to 40 divided buckets. This type is well suited for smaller water volumes with large heads and is often used in alpine power plants. It achieves up to 90% efficiency.

Even though hydroelectric power plants require large volumes of concrete, rebar, stainless steel for pipelines and specialized turbine materials, this resource demand is a singular event at the beginning of the lifetime of such a system. During the PP's operational life, generally, only a few material demands arise given that the service life of turbines and entire dam systems is usually decades rather than years, assuming they receive thorough maintenance. Therefore, from a resource perspective, quantitative demand is not considered a critical issue for hydroelectric power plant systems.

A variation of the storage reservoir is the pumped-storage reservoir, which involves pumping water from a lower reservoir to a higher reservoir. This is commonly done at night, or after periods of strong precipitation or snow melting when surplus energy is available that can be used to operate the pumps and fill the higher reservoirs. As soon as power is required, the potential water energy of the high reservoir is used: water is released back into the lower reservoir, usually to satisfy peak power demand. The material demands are similar to those of the storage type plants, although an additional pipeline system is installed to link the two reservoirs.

For the construction of the turbines, some critical metals are required to produce the necessary stainless steels, however, the quantities are not relevant on a global scale. According to the IEA, hydroelectric power has nearly reached its potential capacity limits in OECD countries (IEA, 2019: x). Growth and the construction of additional hydroelectric power plants are possible in non-OECD countries (FAO, 2019), although estimates about the numbers and sizes of dams and hydroelectric plants are unknown. A critical point is that any dam-type plant interferes with natural water cycles, so while they are positive from an energy gaining point of view, such dams may turn out to be devastating for nature, geophysical conditions and society, both upstream and downstream of the installation.

Tidal power plant

The first tidal PP was La Rance, which began operations in 1966 between St. Malo and Dinard in France. These types of PPs are still rather rare due to the specific geomorphological demands. Several new tidal stream and wave movement systems have evolved and it remains to be seen which systems will persist (if any).

There are three main types of tidal power plants. The first is the classic tidal power plant, which uses the differences in water flows that occur with the tides.

Similar to run-of-river power plants, these also feature dams at suitable locations, which fill up during high tidal movements and are used to feed water through the turbines. During falling and rising tides, the turbines are propelled and produce energy. Gorge-like morphologies are best suited to these PPs, but they are relatively rare. The material demands are similar to those of river type PPs: concrete and rebar for the dam, and corrosion-resistant and stainless steels for the turbines and pipes. One major challenge is coping with the corrosion effects caused by saltwater.

A second, newer type of tidal power plant makes use of sea currents and tidal movements to drive underwater rotors that resemble wind turbines, but on a much smaller scale. There is no dam, but the turbines have to be grounded well by means of special stainless steels and concrete foundations. The primary challenge is to restrict the corrosion of both the foundations and the rotors.

A third type has appeared on the market that uses vertical wave movements. Even though this type of wave movement is rather small and variable, sensitive technology makes it possible to transform this continuous movement into electrical energy. It is still too early to say whether these small energy gains are reason enough to apply this technology on a mass scale. A lot of specialized materials are required to provide the necessary corrosion resistance and capacity to withstand algae and other contamination.

Solar

In solar energy production three principal systems prevail: (1) photovoltaics (PV), where solar energy is turned into electric direct current, (2) solar thermal energy (ST), where a fluid like water is directly heated for immediate use, and (3) concentrated solar power (CSP), where a process fluid is heated by means of mirror arrays to produce electric energy. The primary advantage of CSP is its capability to provide baseload power, as the hot and inert fluid can be processed over a duration of several hours, i.e., day and night. In contrast, PV and ST cannot provide baseload power, because they only function as long as sunlight is available. The potential availability is consequently only half the year, i.e., during daytime hours.

Photovoltaics

A thorough data set about the PV market is available in the PV report provided by the Fraunhofer Institute for Solar Energy Systems (2019). According to the report, PV cell production nearly reached 100 GW in 2018, so that cumulative installed electric solar PV power climbed to 515 GW globally at the end of 2018. PV is thus by far the largest contributor to solar energy (Fraunhofer, 2019). Within PV there are basically three different technological systems: the multicrystalline silicon-based solar cells (multi-Si) that account for about 60% of the PV market, followed by monocrystalline silicon-based cells (Mono-Si), which make up about 35%, and finally thin film PV cells, which have a market share of around 5%. The market is growing fast, with China, Taiwan, and other Asian countries responsible for close to 85% of global production. The record solar cell efficiency is about 26% for mono-Si, 22% for multi-Si, and about 21% for the thin film cells (Fraunhofer, 2019), while marketable cells are a few percent lower. This shows that crystalline PV cells require the greatest quantity of materials, although they only need very few (if any) critical materials.

The very basic functional principle is that solar energy effects an electron flow between two layers within the solar cell, whereby one layer is dominated by positive holes (p-layer) and the other by negative electrons, called an n-type layer. Several more thin layers are contained between the positive and negative electrodes at the front and back end (Ritardo, 2014).

The materials required for the construction of silicon-based cells include aluminum for the back contact, a semiconductor based on silicon and a front contact made of silver. Several materials and alloys can be used for the electrodes, such as silver, gold, aluminum, and others. The thickness of such wafers is in the order of 200 μm. The overall quantitative demand for materials in multicrystalline cells decreased from about 16 g/Wp in 2004 to just over 4 g/Wp in 2017 (Fraunhofer, 2019). Alongside the main material demands, PV systems also require metallic frames, stands, and foundations.

Thin film PV cells contain several critical metals. However the overall quantitative demand for these materials is rather modest as the solar cells are incredibly thin, ranging from some hundred nanometers to just a few micrometers (Lossin, 2010), and their active layers are deposited directly onto a back frame made of glass, plastic or metal. Four categories of thin film cells dominate the market: copper-indium-diselenide (CIS) and copper-indium-gallium-diselenide (CIGS), cadmium-telluride (CdTe), thin-film silicon (a-Si), and gallium-arsenide (GaAs) (Ritardo, 2014). CdTe and CI(G)S cells are the major thin film systems on the market, whereas a-Si and GaAs tend to be used for expensive high-tech and astronautical appliances. Critical aspects of these thin film products are the materials they incorporate. Arsenide, cadmium, gallium, and tellurium show some toxic and thus hazardous characteristics, and therefore have to be handled with care. We can assume they are handled properly during the production process, however these materials pose some challenges when fractured during the operative phase and when they are eventually recycled or deposed of.

It is primarily the materials used in thin film technology that can be seen as critical materials, due to their scarcity, potential toxicity, and the complex manufacturing processes required to achieve the necessary levels of purity. The actual material demand comes from the solar industry, which is mainly located in China. Just three of the top ten producers worldwide are located outside China (Statista, 2019). They include First Solar, the only producer in the United States and the major specialist for CdTe thin film cells.

The future development and potential increase of PV systems, globally, is difficult to predict. Yet PV seems to be a suitable system with large growth potential at the global level. The drawback that PV can only operate during the day when there is sunlight may eventually be solved by the use of batteries to store the energy. With regard to the materials, the production of PV thin film cells could be slowed by the inherent toxicity rather than the availability of critical metals. However, there are still effective silicon-based cells available that can be used as a systemic alternative. Overall, considering the toxicity of the materials required for thin film cells, it is possible that the widespread use of silicon-based cells may have a smaller ecological footprint.

Solar thermal

The direct use of solar energy to heat up water serves rather small appliances. Nevertheless, such systems can support domestic heating systems by raising the base temperature of water for further heating. As such, solar thermal (ST) is primarily used in domestic housing in combination with general heating systems.

The materials demand is comparably modest, as ST rooftop installations only require a few square meters. There are two major types, flat plate solar collectors and heat pipe solar collectors. The flat plate types consist of a back frame made of an aluminum alloy or similar, a transparent top plate, usually some plastics, tubes for the fluids, and insulation materials like polyurethane foams. The tubes are generally made of copper. The heat pipe collectors consist of vacuum tubes that contain copper pipes, insulation materials and a frame, which is usually made out of an aluminum alloy. For both systems, the distance to the domestic heating system has to be rather short so that the stored warmth is not lost. Therefore no large quantities of materials are required. This system will continue to grow, but critical material issues are unlikely.

Concentrated solar power

CSP involves a rather large appliance that produces solar energy on a large scale. It involves mirror systems that are used to concentrate sunlight and heat up a process fluid. This, in turn, is used to produce steam, which drives a generator to produce electricity. CSP's contribution to the global renewable energy supply at the end of 2018 was just under 5.5 GW of installed capacity, with an increase of 550 MW in 2018. Spain is the leading producer of CSP with some 2.5 GW of installed capacity, followed by the United States with approximately 1.5 GW. Most CSP plants have capacities ranging from about 25–150 MW (Helioscsp, 2019). A prominent example of such a plant is Andasol 3, located in southern Spain. The plant has an installed capacity of around 50 MW, which is generated by 205,000 parabolic trough reflectors that are used to heat up thermal oil located in vacuum absorber tubes in the center of the parabolic array. The process fluid is fed to the adjacent power plant which produces the electricity.

The materials demand is determined by both the number of parabolic reflectors, their foundations, and the power plant itself. The parabolic mirrors are made of special curved glass that is 4 mm thick and coated with a thin layer of silver together with several other thin protective coating materials. Each absorber tube consists of a stainless steel tube containing thermal fluid that is encased by an outer glass tube. The challenge is to use materials that can withstand the temperature differences between day and night, as well as during varying solar insolation. A further challenge is ensuring the precise orientation of the parabolic mirrors in relation to the sun. Here a strong foundation for the mirrors is required, as well as a precise system for aligning the mirrors to obtain the maximum yield (Solar Millennium, 2008).

Solar power tower plants are another type of CSP system, whereby a process fluid in a centrally located tower is heated by aligned mirrors or heliostats. The principle is the same as for the Andasol plant. There are several more systems that all concentrate solar energy using a suitable method before turning it into electricity.

There is certainly growth potential for such base load-capable systems. However, CSP plants are expensive and optimal conditions usually exist in places where there is less demand. This means that large grid systems are required to make use of such plants.

Wind energy

Wind turbine generators (WTG) convert wind energy from mechanical into electrical energy. The main WTGs used nowadays are the horizontal axis drive train, three-blade wind turbines. The major components are the rotor and hub, the nacelle that houses the drive train system (DTS) and the energy transformation units, the tower that hosts the rotor and nacelle, and the foundation that provides the necessary stability. The type of conversion system, the DTS, makes up a great deal of the nacelle's weight and material requirements. The required height, rotor size, and stability of the tower are determined by the weight of the nacelle, the prevailing wind speeds and gusts, and the hardness of the ground.

With regard to the materials, it is necessary to differentiate between the conversion systems. Two major conversion types dominate the market. Asynchronous generators (ASG) have a drive with a gearbox that is typically connected to a doubly-fed asynchronous induction generator (DFIG). This design principle is relatively simple and thus reliable, inexpensive, well developed, and widely used (De Lara Garcia, 2013). It is the standard onshore WTG type. The other major system with an increasing market share is the synchronous generator (SG) type, which is a direct drive (DD) system. The transformation and energy generation occurs directly by means of a magnet system. These DD systems can either use an electrically excited magnet (like the WTGs produced by the German company ENERCON) or a permanent magnet system. The permanent magnet generator (PMG) type is especially well suited for offshore use, as there is no gearbox and thus fewer parts, which reduces the need for maintenance. Whereas onshore WTGs can easily be accessed at any time, offshore platforms require more effort to access; they may become impossible to access during storms, so excellent reliability is paramount for offshore WTGs. These PMG-based turbines use permanent NdFeB magnets that contain the rare earth element (REE) neodymium and possibly also praseodymium, dysprosium, terbium, and gadolinium. As China has a monopoly on production with more than 80% of global supply, these materials are considered critical. The exact quantity and composition of the permanent magnets required for WTGs varies and may change from turbine to turbine. According to the molar mass conversion, the share of rare earth elements in an $Nd_2Fe_{14}B$ magnet is around 27% neodymium, whereby praseodymium and dysprosium may substitute or add a few percent to the given value. For WTGs of the PMG type, the rule of thumb is 600 kg of magnet per MW of installed power (Glöser-Chahoud, Kühn, & Tercero, 2016; Zepf, 2012). Of this, about 170 kg/MW are neodymium and possibly about 24 kg/MW (4%) are dysprosium. A standard 5 MW PMG WTG thereby has a magnet weight of 3 tons, of which about 840 kg are neodymium and 120 kg are dysprosium.

The rotor blades are considered to be the most expensive components of a WTG. The primary materials are glass and carbon fibers, which are manufactured as two faces joined together and stiffened by a web or box structure (Mishnaevrsky et al., 2017). The rotors are highly critical parts of a WTG, as they experience extremely high speeds and forces (bending and compression) toward the wing tips. In addition, they are faced with environmental conditions like dust, rain, and hail. Minor imperfections, cracks, bondlines, bubbles, etc. may lead to structural damage, so even the smallest production faults have to be avoided and any damage must be immediately repaired.

ASG-type WTGs contain a gearbox consisting of stainless steel and other special steels. The complexity of such gearboxes mean they potentially require greater maintenance; however, the technology is well understood and has been optimized for decades now. Consequently, these mechanical systems still provide the bulk of wind energy, primarily onshore, but there are also offshore WTGs of the ASG type.

The towers are mostly tubular steel constructions that also contain special steels to guarantee stability. The amount of concrete and rebar required for the foundation depends on the ground structure, the condition and stiffness of the soil, and the prevailing wind and turbulences. Svensson (2010) compares various foundation systems and reports that the pile length for standard WTGs may be as much as 3500 m, which determines the major quantitative part of the material demand. A wide spectrum of systems are used for offshore foundations; the most common are monopoles, tripods, and jacket systems. Resistance to corrosion caused by saltwater is paramount for all of these foundations and is provided by stainless steels and tailored maintenance schedules. Mono pile foundations are tubular steel structures up to 60 m long with a diameter of between 4 and 8 m. They are suitable for shallow water depths of around 25–40 m. A jacket foundation can be used in deeper waters ranging from 30 to 80 m; they are stiffer and therefore less susceptible to wave and current load factors (Zhang, Fowai, & Sun, 2016).

Ultimately, "there may not exist the best wind turbine generator technology to tick all the boxes" (Cao, Xie, & Tan, 2012). Instead, the decision will always be a compromise that takes into consideration the cost of the system, i.e., material and operational cost, the environmental and geographical conditions, and the economic wind market aspects. Table 4.2 provides an overview of the components and system differences. For example, for more details refer to Schubel and Crossley (2012).

The Global Wind Energy Council estimated there were a total of 341,000 WTGs installed globally at the end of 2016 (GWEC, 2018) with a realistic average WTG capacity of 1.5 MW. While there is no current data for offshore wind turbines available, there are likely some 8000 offshore WTGs in operation (23 GW of documented

Table 4.2 Wind energy—Technical and materials-related data.

Wind turbine generator	Type	Senvion MM82	Senvion 3.M122	Senvion 6.2M126	Adwen AD5-116 (M5000)[a]	Siemens SWT 3.1-101
General	Nominal power	2 MW	3 MW	6.2 MW	5 MW	3.2 MW
	Drive train system DTS	ASG DFIG	ASG DFIG	ASG DFIG	Hybrid	DD PMG
	Hub height	58.5–80 m	136–139 m	95–117 m	90 m	74.5–94 m
Rotor blade	Length	40 m	59.8 m	61.5 m	116 m diameter	49 m
	Weight	6 tons	15 tons	25.5 tons	unknown	20 tons
Hub	Weight	17 tons	25 tons	82 tons	109 tons incl. rotor	32 tons
Nacelle	Weight	70 tons	58 tons (without DTS)	350 tons (without DTS)	200 tons	78 tons
Data source		Senvion (2014)	Senvion (2014)	Senvion (2014)	Ventus (2015)	Siemens (2016)

[a] Weight of steel in tripod: 700 tons; weight of steel in tower and nacelle: 300 tons.
ASG, asynchronous generator; DD, direct drive; DFIG, doubly-fed induction generator; DTS, drive train system; IEC, International Electrotechnical Commission; MW, megawatt; PMG, permanent magnet generator; WTG, wind turbine generator.

installed capacity in 2016 divided by an estimated average capacity of 3 MW per offshore WTGs).

In 2001, when the first three-blade WTGs entered the market, the total global installed wind energy capacity was 23 GW. This increased to 600 GW by the end of 2018. It is estimated that installed capacity of 800 GW will be reached by the end of 2022 (GWEC, 2018, 2019). In 2017 wind energy became a fully commercialized and generally unsubsidized energy technology that has proven economically competitive, so the estimates for 2022 seem realistic.

To be clear, not all WTGs are of the PMG type, and it will not be possible to produce this forecast capacity exclusively with PMG-based wind turbine generators. However, there were and still are proven alternatives that require fewer critical materials, which will guarantee future growth in this sector.

Geothermal energy

Geothermal energy makes use of the temperature of the Earth. According to Auer (2010), 50%–70% of the inherent temperature of the Earth was stored during the Earth's formation and the remaining 30%–50% originates from the natural decay of radioactive isotopes. This continuous source of heat is therefore considered to be one of the most reliable and "ever ranging" systems. Yet geothermal energy is still a niche industry, as the global installed capacity was little more than 13 GW in 2018 (IRENA, 2019).

Typically, two main geological areas are of interest when it comes to geothermal energy. The first includes heat anomalies that are often linked to the active magmatism found near plate tectonic boundaries and geological mantle hot spot areas (Ni, Velasquez, & Gonzalez, 2016). The use of such occurrences is thus limited to a few areas on Earth and they do not represent a general option. These areas are considered to be high-enthalpy resources with temperatures of around 180–200°C and more (Auer, 2010; Ni et al., 2016). This energy is primarily suitable for electricity generation. The second type of geological area is without active magmatism. These low-enthalpy resources are intermediate (100–180°C) and low-temperature resources (less than 100°C), which can be found in continental settings. Here the geothermal gradient can indicate how warm or hot it will become with increasing depth. As a rule of thumb, there is a mean temperature increase of 3°C per 100 m of depth in continental crust layers.

The applications of these two resources are in principle twofold: either *directly* for heating and cooling, primarily in domestic applications, or *indirectly* in electric power plants (called binary systems). At present these power plants are mostly of the dry steam or flash plant types. In addition, binary cycle systems are becoming mature; they make it possible to use intermediate temperature resources in heat exchanger systems to produce electricity.

The differences between direct and indirect geothermal energy use is also reflected in the materials demand. The construction of a geothermal power plant primarily requires concrete, rebar for the construction of the buildings, and specialized steels for the processing and transformation of heat into electricity. Heat exchange systems also require various specialized steels and polymers, regardless of whether they are used in large-scale power plants or in domestic housing. All geothermal systems need a pipe system to feed the hot liquids from the depths of the Earth to the surface and back again. The depth and number of boreholes and the temperature of the hot fluids determines the quality and quantity of the materials required. For example, one of the largest geothermal power plants in the world, the Hellisheidi geothermal power plant in Iceland, provides 303 MW of electricity and 103 MW of thermal power. The plant construction began in 2004 and it was fully operational by 2011. In total, 40 of the 57 drilled boreholes are in use today, of which most have a depth of 2000 m and a total length of 3000 m (Reykjavik Energy, n.d.). The larger plants are usually complexes like the Geysers Complex in California, which has an installed capacity of 1520 MW and 22 power plants, or the Laradello Complex in Italy, which has an installed capacity of 720 MW and 34 plants (WorldAtlas, 2019).

Geothermal energy can also be used to supply the heating systems in domestic housing. Depending on the ground conditions, a pipe system is installed either inside a hole that is a few decameters depth, or as a flat collector or spiral collector system that is only a few meters underground. These pipes are made of stainless steels or plastic tubes and are part of a heat exchanger system. System fluids are circulated through these pipes in order to extract the heat from the Earth.

Overall, the materials demand for geothermal systems is rather modest. As only a few locations on Earth are suitable for large geothermal power plants, no materials shortage is expected. In domestic applications the required depth for the probes is relatively small, so that even with increasing global demand for such systems, there is unlikely to be excessive demand for materials.

Conclusion: Materials demand for renewable energies

The previous paragraphs showed that for biomass, hydro and geothermal energy systems there is no demand for critical materials, as long as special steels, rebar and the sand used to make concrete are not considered critical. With regard to biomass, the way in which biomass fuels are produced may, however, attribute some overall negative factors to its otherwise clean and base load-capable energy. A range of critical materials is required for the production of thin film PV cells, however, their global share of the PV market is only about 5%, which shows that systemic alternatives exist (primarily silicon-based PV cells) (Fraunhofer, 2019). A similar situation prevails in the wind energy sector. Only direct drive systems that feature permanent magnets use some REEs. The main REE found in these magnets is neodymium, which is considered critical not least because such magnets are also widely used in electromobility and industrial appliances. Again, here the share of these types of WTG is around 20% of the global market, which shows that systemic alternatives exist.

Due to this vast complexity in clean energy systems, it is, therefore, advisable to have an idea of the supply chains so that the respective trade links can be established and diversified. The US Geological Survey and many other surveys provide regular reports on global mining. This data provides an initial overview of the many countries, companies, and actors producing raw materials. Some dedicated papers and studies elaborate on the potential future demand, such as the studies *Rohstoffe für Zukunftstechnologien 2016* (Marscheider-Weidemann et al., 2016) and *Metal demand for Renewable Electricity Generation in the Netherlands* (Van Exter, Bosch, Schipper, Sprecher, & Kleijn, 2018). All the calculations are based on a range of assumptions that reflect several scenarios; they do not reflect reality, but rather the likelihood of different extremes or parameter sets that may be caused by future developments.

But irrespective of any definitions of criticality and resources, the vast number of renewable energy systems entails demand for an even larger variety of materials. Very often the major quantitative demand is for base metals and rather conventional materials such as concrete, rebar, and stainless steel. Special materials demand is certainly also an issue and several elements are required to achieve desired functionalities, such as the REE neodymium, which is vital for the manufacture of extremely strong permanent magnets. Unlike iron-based magnets, these magnets can be used in large WTGs. However, there are systemic alternatives. There are several types of WTG currently in use and only one of them features neodymium-based permanent magnets as the main energy conversion system. A similar situation exists in relation to solar energy production. PV systems can use crystalline silicon-based solar cells, or thin film, for example, gallium-based cells. Even though GaAs cells are more efficient (and more expensive), it is still possible to produce solar energy without gallium.

In addition to the systemic alternatives available for each type of renewable energy, major differences in size, environmental, sociocultural, and geographic conditions also result in a wide range of quantitative material and qualitative resource demands. This means it is not possible to calculate the average material and resource demands in a sound scientific or realistic manner, and thus the results of most studies show numbers that only partially reflect reality. Too many factors have to be considered and, especially on a global level, it seems impossible to predict how renewable systems will develop, what technologies are planned for their implementation, and thus the materials demand. It seems even more impossible to suggest which energy system will have the best results. One energy system may turn out to be the most suitable in country A due to geographical (e.g., long insolation), geophysical (e.g., persisting strong winds), social (e.g., acceptance of technology), economic (e.g., available expertise and materials), or political reasons (e.g., subsidies or regulatory framework), whereas the same system may be completely impractical in country B.

Considering all these interdependencies and many variants, it should be obvious that there is no one ideal system that is suitable for everything. Rather, for each energy demand, an assessment has to be performed in order to weigh up all the options and their respective pros and cons so that the most suitable system for the given situation can be selected. As a consequence, it is possible that a preferred renewable energy system has to be refused, because the necessary critical materials cannot be attained. In such cases, the second best choice may be the most suitable one.

For the end user or customer, the topic of critical materials is one aspect to be considered early on. Taking critical minerals into account can help end users choose the most suitable renewable energy system for their individual situation. For producers and manufacturers of renewable energy systems, the question of critical materials provides an opportunity to develop strategies and systemic alternatives early on to counter potential supply risks.

Past experience has shown that whenever materials have become scarce or unattainable, more research has been undertaken to improve material efficiency, find new supplies (i.e., deposits), and develop material substitutes and systemic alternatives. This may take time, effort, and a lot of money, but if the function enabled by those materials is truly necessary, then this pathway is a probable one. It is also likely that if raw materials become really scarce and thus expensive, recycling will eventually become economical and competitive compared to primary raw materials. Recycling could provide a secondary supply and thus relieve the supply pressure.

References

Achzet, B., Reller, A., Zepf, V., Ashfield, M., Simmons, J., & Rennie, C. (2011). *Materials critical to the energy industry. An introduction.* London.

Adachi, G., Imanaka, N., & Tamura, S. (2010). Research trends in rare earths: A preliminary analysis. *Journal of Rare Earths, 28*(6), 843–846.

Auer, J. (2010). *Geothermal energy.* Frankfurt am Main: Deutsche Bank Research.

Böschen, S., Soentgen, J., & Reller, A. (2004). Stoffgeschichten—Eine neue Perspektive für transdisziplinäre Umweltforschung. *Gaia, 13*(1), 19–25.

Cao, W., Xie, Y., & Tan, Z. (2012). *Wind turbine generator technologies.* https://doi.org/10.5772/51780.

De Lara Garcia, J. P. S. (2013). *Wind turbine database: Modelling and analysis with focus on upscaling.* http://publications.lib.chalmers.se/records/fulltext/179591/179591.pdf. (Accessed July 12, 2019).

European Commission. (2018). *Report on critical raw materials and the circular economy.* Brussels. https://doi.org/10.2873/167813.

FAO. (2019). *Aquastat. Dams. Geo-referenced database.* http://www.fao.org/nr/water/aquastat/dams/index.stm. (Accessed September 28, 2019).

Fortier, S. M., Nassar, N. T., Lederer, G. W., Brainard, J., Gambogi, J., & McCullough, E. A. (2018). *Draft critical mineral list—Summary of methodology and background information—U.S. geological survey technical input document in response to secretarial order no. 3359.* U.S. geological survey open-file report 2018-1021. Reston. https://doi.org/10.3133/ofr20181021.

Fraunhofer. (2019). *Photovoltaics report.* Freiburg: Fraunhofer ISE.

Glencore. (2018). *Responsibly sourcing the commodities for everyday life. Annual report 2018.* Baar.

Glöser-Chahoud, S., Kühn, A., & Tercero, E. L. A. (2016). *Globale Verwendungsstrukturen der Magnetwerkstoffe Neodym und Dysprosium.* Karlsruhe: Fraunhofer ISI.

GWEC. (2018). *Global wind report. Annual market update, 2017.*

GWEC. (2019). *51.3 GW of global wind capacity installed in 2018.* Press release, 26 February 2019.

Helioscsp. (2019). *Concentrated solar power increasing cumulative global capacity more than 11% to just under 5.5 GW in 2018.* Press release, 18 June 2019 http://helioscsp.com/concentrated-solar-power-increasing-cumulative-global-capacity-more-than-11-to-just-under-5-5-gw-in-2018. (Accessed July 12, 2019).

IEA. (2019). *Renewables information: Overview* (2019 edition). https://webstore.iea.org/renewables-information-2019-overview. (Accessed September 28, 2019).

IRENA. (2019). *Renewable energy statistics 2019.* Abu Dhabi: The International Renewable Energy Agency.

Krausmann, F., Gingrich, S., Eisenmenger, N., Erb, K. H., Haberl, H., Fischer- Kowalski, M. (2009). Growth in global materials use, GDP and population during the 20th century. *Ecological Economics, 68*(10), 2696–2705.

Lossin, A. (2010). Sondermetalle in Solarzellen. In *GDMB. Sondermetalle und Edelmetalle. Vorträge beim 44. Metallurgischen Seminar. Heft 121 der Schriftenreihe der GDMB Gesellschaft für Bergbau, Metallurgie, Rohstoff- und Umwelttechnik e.V.*

Marscheider-Weidemann, F., Langkau, S., Hummen, T., Erdmann, L., Tercero Espinoza, L., Angerer, G., et al. (2016). Rohstoffe für Zukunftstechnologien 2016. In *DERA Rohstoffinformationen 28.* Berlin.

Mishnaevrsky, L., Jr., Branner, K., Petersen, H. N., Beason, J., McGugan, M., & Sørensen, B. F. (2017). *Materials for wind turbine blades: An overview.* https://doi.org/10.3390/ma10111285.

Ng, E. (2019). *China's war chest of rare earth patents give an insight into total domination of the industry.* South China morning post, 20 July 2019. https://www.scmp.com/business/companies/article/3019290/chinas-war-chest-rare-earth-patents-give-insight-total. (Accessed September 28, 2019).

Ni, O., Velasquez, R., & Gonzalez, G. (2016). Sustainability assessment, case of study: Geothermal power plant. In *14th LACCEI International Multi-Conference for Engineering, Education, and Technology: "Engineering Innovations for Global Sustainability", 20-22 July 2016, San José, Costa Rica.*

Niu, X. (2016). Key technologies of the hydraulic structures of the three Gorges project. *Engineering, 2*(2016), 340–349. https://doi.org/10.1016/J.ENG.2016.03.006.

NRC. (2008). *Minerals, critical minerals, and the U.S. economy.* National Research Council of the National Academies Washington D.C.

Reykjavik Energy (n.d.) Geothermal power plants. Brochure. www.oi.is (Accessed 12 July 2019).

Ritardo, D. T. (2014). Photovoltaic systems. In S. A. Kalogirou (Ed.), *Solar energy engineering.* (2nd ed., pp. 481–540). https://doi.org/10.1016/B978-0-12-397270-5.00009-1.

Schubel, P. J., & Crossley, R. J. (2012). *Wind turbine blade design.* https://doi.org/10.3390/en5093425.

Senvion. (2014). *Product portfolio overview. Technical data.* https://www.senvion.com/global/en/products-services/. (Accessed July 12, 2019).

Siemens. (2016). *Wind turbine SWT 3.2-101. Data sheet.* https://www.wind-turbine-models.com/turbines/964-siemens-swt-3.2-101. (Accessed July 12, 2019).

Solar Millennium. (2008). Die Parabolrinnen-Kraftwerke Andasol 1 bis 3. In *Die größten Solarkraftwerke der Welt; Premiere der Technologie in Europa.* Technical brochure. https://iam.innogy.com/-/media/innogy/documents/ueber-innogy/innogy-Innovation-und-Technik/PDF-solarkraftwerk_andasol-weitere_informationen.pdf. (Accessed July 11, 2019).

Statista. (2019). *Photovoltaik in Deutschland.* https://de.statista.com/statistik/studie/id/6548/dokument/solarenergie-statista-dossier/. (Accessed July 12, 2019).

Svensson, H. (2010). *Design of foundations for wind turbines*. Master's Dissertation, Lund.

UNEP. (2013). *Metal recycling: Opportunities, limits, infrastructure*. A Report of the Working Group on the Global Metal Flows to the International Resource Panel.

US DoE. (2010). *U.S. Department of Energy. Critical materials strategy*.

USBR. (2019). *Upper Colorado Region. Glen Canyon Unit. Quick Facts*. United States Bureau of Reclamation. https://www.usbr.gov/uc/rm/crsp/gc/. (Accessed July 12, 2019).

USGS. (2019a). *Mineral commodity summaries 2019*. Reston. https://doi.org/10.3133/70202434.

USGS. (2019b). *Appendices 2019*. https://prd-wret.s3-us-west-2.amazonaws.com/assets/palladium/production/s3fs-public/atoms/files/mcsapp2019.pdf. (Accessed July 12, 2019).

Van Exter, P., Bosch, S., Schipper, B., Sprecher, B., & Kleijn, R. (2018). *Metal demand for renewable electricity generation in the netherlands*. Navigating a complex supply chain. Leiden.

Ventus, A. (2015). *Fact sheet*. https://www.alpha-ventus.de/fileadmin/Dateien/publikationen/av_Factsheet_Engl_2016.pdf. (Accessed July 12, 2019).

Verbund. (2019). *Our power plants*. https://www.verbund.com/en-at/about-verbund/power-plants/our-power-plants. (Accessed July 12, 2019).

World Bank. (2019). Climate smart mining. In *Minerals for climate action. Infographic*. https://www.worldbank.org/en/topic/extractiveindustries/brief/climate-smart-mining-minerals-for-climate-action. (Accessed July 12, 2019).

WorldAtlas. (2019). *Largest geothermal power plants in the world*. https://www.worldatlas.com/articles/largest-geothermal-power-plants-in-the-world.html. (Accessed July 12, 2019).

Yang, Y., Walton, A., Sheridan, R., Güth, K., Gauß, R., Gutfleisch, O., et al. (2016). *REE recovery from end-of-life NdFeB permanent magnet scrap: A critical review*. https://doi.org/10.1007/s40831-016-0090-4.

Zepf, V. (2012). Rare Earth Elements. In *A new approach to the nexus of supply, demand and use. Exemplified along the use of neodymium in permanent magnets*. Berlin/Heidelberg: Springer.

Zepf, V., Reller, A., Rennie, C., Ashfield, M., & Simmons, J. (2014). *Materials critical to the energy industry. An introduction* (2nd ed.). London.

Zhang, J., Fowai, I., & Sun, K. (2016). *A glance at offshore wind turbine foundation structures*. https://doi.org/10.21278/brod67204.

Stationary battery systems: Future challenges regarding resources, recycling, and sustainability

5

Marcel Weil[a,b], Jens Peters[b], and Manuel Baumann[a]

[a]*Institute for Technology Assessment and Systems Analysis (ITAS)/Karlsruhe Institute of Technology, Karlsruhe, Germany*
[b]*Helmholtz Institute Ulm for Electrochemical Energy Storage (HIU), Ulm, Germany*

Background

A total of 195 countries signed a global climate deal at the Paris climate conference (COP21) in December 2015. The main aim of the agreement is to avoid uncontrolled climate change and to ensure the survival of mankind by keeping the global temperature increase below (at least) 2°C, and preferably below 1.5°C compared to preindustrial levels (Masson-Delmotte et al., 2019).

To achieve such ambitious goals, there is an urgent need for worldwide action to radically transform our present society into a low-carbon society that releases minimum emissions of carbon dioxide and other anthropogenic greenhouse gases into the atmosphere (Peake, 2012). To realize a low-carbon society, all sectors that cause significant carbon dioxide and greenhouse gas emissions need to fundamentally change. The energy sector is one such sector, and the ongoing worldwide energy system transformation is fostering the use of regenerative energy sources such as wind and solar power, and phasing out conventional fossil fuel-based power generation. However, a high share of regenerative energy sources within the grid leads to increasing power supply fluctuation, which causes significant challenges for the continuous provision of energy and management of dynamically changing loads (Weil & Tübke, 2015). Flexibility options like energy storage make it possible to integrate large quantities of renewable energy sources into the energy system and can ensure a resilient and secure energy supply for industry and society as a whole in the future.

Among the existing short- to mid-term storage options, pumped hydro storage (PHES) and batteries are currently the dominant technologies, competing with others such as diabatic and adiabatic compressed air energy storage (CAES and ACAES). Due to historical reasons and their low costs, PHES facilities are the biggest source of global energy storage capacities. However, their expansion is limited due to geographical and geological conditions, as well as societal acceptance issues. In contrast,

The Material Basis of Energy Transitions. https://doi.org/10.1016/B978-0-12-819534-5.00005-2

battery technologies are currently experiencing the highest growth rates in the field of energy storage systems.

Different kinds of batteries are used for grid energy storage worldwide, with lithium-ion batteries (LIB) being the dominating cell technology (CNESA, 2018). LIBs were the technology of choice in 85% of the stationary energy storage projects commissioned in 2016, and their share further increased to 90% in 2017 (CNESA, 2018). Lead-acid batteries, sodium-sulfur (NaS) batteries, and vanadium redox flow batteries (VRFB) play only minor roles within the stationary battery sector nowadays (CNESA, 2018). Thus LIBs continue to increase their market dominance, but unfortunately require raw materials that are considered critical, potentially critical, or vulnerable (Weil & Ziemann, 2014), such as Co, Li, Ni, or graphite.

Future demand for stationary energy storage

The energy system transformation is essentially based on two regenerative energy sources: wind and sunlight. Due to the strong dependence of renewable energy conversion plants on wind and solar radiation and the associated fluctuation in energy production, batteries play an important role in providing grid stability and safeguarding energy supply on an hourly to daily basis. In addition, batteries are used at the household level (mainly rooftop photovoltaics) to increase energy self-consumption, but sometimes these batteries even feed electricity into the grid. In such cases, the house owner (or tenant) not only consumes energy but also produces energy, making them a "prosumer."

On behalf of the Energy Watch Group (EWG), a dynamic global model of the energy system was developed in order to estimate the implications of a transition to 100% renewable energy sources by 2050 (Ram et al., 2019). Of course, a model with such ambitious goals can be considered as an extreme future scenario, but at the same time, it addresses the ambitious goals of the COP21 agreement. The model makes it possible to quantify aspects such as the theoretical cumulative energy storage demand in the years 2030 and 2050 (see Fig. 5.1). The results reveal a tremendous need for energy storage units. The total demand (for batteries, PHES, and ACAES) amounts to nearly 20,000 GWh in 2030 and over 90,000 GWh in 2050. The battery storage requirements alone (grid and prosumer) are forecast to reach approximately 8400 GWh in 2030 and 74,000 GWh in 2050. Based on these numbers, it is possible to estimate the potential resource demands involved in producing and replacing all the stationary batteries that will be required by 2050.

Potential lithium-based electrochemical energy storage options

As previously mentioned, the present battery market for stationary applications is dominated by lithium-ion batteries (LIB) (CNESA, 2018). There are many different

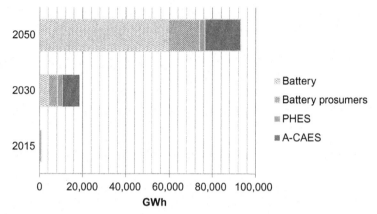

FIG. 5.1

Energy storage demand for 2030 and 2050: PHES (pumped hydroelectric energy storage) and A-CAES (adiabatic compressed air energy storage).

Credit: Data from Ram, M., Bogdanoy, D., Aghahosseini, A., Gulagi, A., Oyewo, A. S., Child, M., Caldera, U., Sadovskaia, K., Farfan, J., Barbosa, L. S. N. S., Fasihi, M., Khalili, S., Dalheimer, B., Gruber, G., Traber, T., De Caluwe, F., Fell, H.-J., Breyer, C., 2019. Global energy system based on 100% renewable energy—Power, heat, transport and desalination sectors. Study by Lappeenranta University of Technology and Energy Watch Group: Lappenranta, Berlin.

LIB technologies available on the market, which have fundamentally different cathode chemistries and therefore also contain very different types and amounts of metals (Table 5.1). On the anode side, the vast majority of LIBs available on the market rely on natural graphite, which is not addressed in this study. Due to their varying technical performance (e.g., energy density, power density, safety, robustness, cycle life) and different costs, the individual LIB chemistries are suited to (and used in) different application fields (Zubi, Dufo-López, Carvalho, & Pasaoglu, 2018). The high energy

Table 5.1 Metal content of different lithium-ion batteries, g/kWh.

	Li	Co	Ni	Mn
LIB-LCO	113	959	0	0
LIB-NCA	112	143	759	0
LIB-NMC-111	139	394	392	367
LIB-NMC-622	126	214	641	200
LIB-NMC-811	111	94	750	88
LIB-LFP	338	0	9	0

Based on (Olivetti, E. A., Ceder, G., Gaustad, G. G., Fu, X., 2017. Lithium-ion battery supply chain considerations: Analysis of potential bottlenecks in critical metals. Joule, 1, 229–243. doi:10.1016/j. joule.2017.08.019) and (Majeau-Bettez, G., Hawkins, T. R., Strømman, A. H., 2011. Life cycle environmental assessment of lithium-ion and nickel metal hydride batteries for plug-in hybrid and battery electric vehicles. Environmental Science & Technology, 45, 4548–4554. doi:10.1021/ es103607c).

density of lithium cobalt oxide (LIB-LCO) has made it the preferred chemical compound for cell phones, tablets, and laptops. LIB-LCO was also used in cars like the Tesla Roadster and the Smart Fortwo electric drive (Miao, Hynan, von Jouanne, & Yokochi, 2019), and airplanes like the Boing 787 Dreamliner (for auxiliary startup and backup power during flights) (Zubi et al., 2018) but was soon replaced due to some serious safety issues and battery failure during use.

Lithium nickel cobalt aluminum oxide batteries (LIB-NCA) have good energy and power density and a long cycle life. They are used in electric vehicles (especially in most Tesla cars), and research suggests these batteries could be pooled and integrated into electricity grids to perform load shifting and provide backup power (Zubi et al., 2018).

Nickel manganese cobalt batteries (LIB-NMC) can be designed for both high-energy and high-power applications. Thanks to their good technical performance and thermal behavior, they are used in many applications including electronics and electric vehicles (Nissan Leaf, Chevy Volt, and BMW i3) (Miao et al., 2019). LIB-NMC is also used for different stationary applications (Hesse, Schimpe, Kucevic, & Jossen, 2017). In our investigation, we consider three different types: NMC-111, NMC-622, and NMC-811 (Table 5.1). The numbers denote the ratio of Ni, Co, and Mn on a mole fraction basis, thus NMC-111 is a high Co, low Ni system, whereas NMC-811 is a low Co, high Ni system.

Lithium-iron-phosphate batteries (LIB-LFP) do not require the use of expensive metals like nickel and cobalt while offering sufficiently good electrochemical performance with low resistance, a wide operating temperature range, and a long cycle life (Miao et al., 2019). On the other hand, LFP has significantly lower energy density than NCA, LCO, or NMC systems. The Chinese manufacturer BYD uses large quantities of LIB-LFP for the production of electric buses (Weil, Peters, & Baumann, 2019), but LFP is less commonly considered for the production of electric vehicles (EVs). LFP is also used in many stationary energy storage projects for different grid services (Hesse et al., 2017; Tsiropoulos, Tarvydas, & Lebedeva, 2018).

Calculation of the potential resource demand for batteries
Approach, assumptions, and scenarios

The potential resource demand for a full energy system transformation is calculated on the basis of the previously outlined demand for stationary battery storage by 2050 (Fig. 5.1) and the metal composition of different LIBs (Table 5.1). Although the future market will most probably feature a mixture of different battery types, the potential 2050 scenarios are each calculated with one single LIB type. This makes it easier to identify the differences in resource demands between the individual cell chemistries. The service life of a battery system for a stationary application is estimated to be 20 years; battery cells with shorter service lives are assumed to be replaced during this time, adding to the total resource demand. To work out the effects of battery recycling on the total resource demand, two recycling rates (RR) are considered: 50% and 100%. The recycling rate not only comprises the recycling efficiency of the recycling process itself, but also the collection of the old batteries and some pretreatment. For some metals, like Li, a 50% RR is already a great challenge, because at present Li is often not recovered during the current recycling process. In contrast, the recycling efficiency for metals like Co and Ni is already quite high and could be close to 90% (Mohr, Weil, Peters, & Wang, 2020). It is important to point out that the results exclude metal usage in other sectors, and only take into account the demand for stationary batteries. Obviously, the mobile application of batteries in electric vehicles (EV) will contribute significantly to the total battery demand (Vaalma, Buchholz, Weil, & Passerini, 2018). On the other hand, the demand for batteries for stationary applications would decrease if EV batteries were used for short-term stationary services (vehicle to grid concept), or if old EV batteries were reused for stationary applications (second use). Regarding the materials demand of batteries for EV applications, please refer to previous studies (Weil, Ziemann, & Peters, 2018; Ziemann, Müller, Schebek, & Weil, 2018).

Scenario for LCO

Fig. 5.2 shows the cumulative demand for Li and Co by 2050 for stationary applications involving an LCO-type battery, compared with the known lithium reserves and resources in 2018 (USGS, 2019). "Reserves" describes the known amount of a resource that can be extracted economically, while "resources" refers to the amount of a resource that could theoretically be obtained using existing technology. In the case of Li, by 2050 more than 60% of the presently known reserves will have been consumed, but less than 15% of the known resources. In contrast, by the same year, the demand for Co would be 3 times higher than the known resources. Even two-thirds of the 120 million tons of submarine Co resources (USGS, 2019) would have also been consumed. Submarine Co resources have been identified in manganese nodules and crusts on the floor of the Atlantic, Indian, and Pacific Oceans (USGS, 2019).

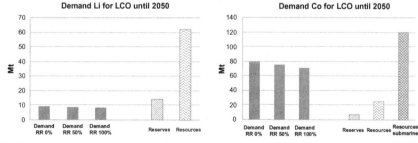

FIG. 5.2

Calculated cumulative Li and Co demands for LCO systems by 2050 in Mt with theoretical recovery rates of 50% and 100%, compared with reserve and resource data.

Credit: Reserve and resource data from USGS, 2019. USGS mineral commodities survey, US geological surveys. *US-DOI.*

Deep-sea mining concepts and mining equipment have already been developed (van Wijk, 2018), but seabed mineral extraction in the deep sea is still a very controversial topic due to the anticipated (and potentially significant) impacts on marine ecosystems (Kaikkonen, Venesjärvi, Nygård, & Kuikka, 2018).

Recycling the metals in batteries reduces the demand for primary resources. To evaluate the effects of recycling, we have considered two theoretical recovery rates: 50% and 100%. However, even with a theoretical recovery rate of 100%, the reduction in primary raw material demand appears rather low for Li and Co. This is an effect of the batteries' comparably long service lives and the high growth rates expected between now and 2050. As such, the stream of recovered material (even in the case of a fully circular economy) would only cover a marginal share of the annual resource demand.

Scenario for NMC 111

Fig. 5.3 illustrates the cumulative demand for Li, Co, Ni, and Mn for the manufacture of NMC111-type batteries by 2050 and is compared with the known reserves and resources data from 2018 (USGS, 2019). In the case of Li, the expected demand is even higher than for LCO systems. If demand is met, more than 80% of the presently known reserves will have been consumed by 2050, but less than 20% of the known

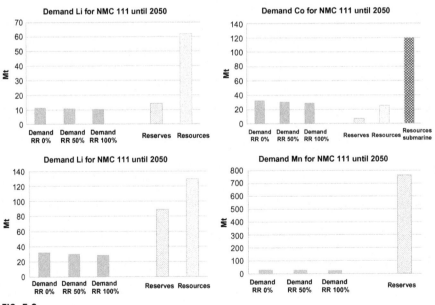

FIG. 5.3

Calculated cumulative Li, Co, Ni, and Mn demands for NMC 111 systems by 2050 in Mt with theoretical recovery rates of 50% and 100%, compared with reserve and resource data.

Credit: Reserve and resource data from USGS, 2019. USGS mineral commodities survey, US geological surveys. *US-DOI.*

resources. The Co demand by the same year would be higher than the 25 million tons of known resources—even with the 100% recycling rate—and would involve the consumption of approximately one-quarter of the known submarine resources.

In the case of Ni, by 2050 approximately one-third of the known reserves and less than one-quarter of the known resources would have been consumed. Mn can be considered the least critical constituent, as less than 10% of the known reserves will be consumed by 2050. The Mn resource data is not displayed in Fig. 5.3, because of the relatively good availability of Mn reserves.

Scenario for NMC 622

For the NMC 622-type battery, the cumulative demand by 2050 for Li, Co, Ni, and Mn is displayed in Fig. 5.4, and compared with the known reserves and resources data from 2018 (USGS, 2019). The most interesting changes in comparison to NMC 111 batteries relate to Co and Ni. In the case of Co, the demand by 2050 is more than one-third smaller and accounts for 18 million tons without 100% recycling, and 16 million tons with 100% recycling. The known reserves appear rather small in comparison, at approximately 7 million tons. Also, roughly two-thirds of the known resources would be consumed by 2050. The Ni demand for NMC 622 would be

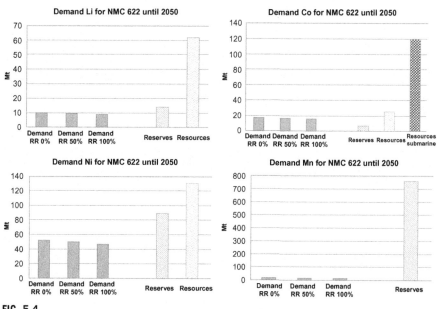

FIG. 5.4

Calculated cumulative Li, Co, Ni, and Mn demand for NMC 622 systems by 2050 in Mt with theoretical recovery rates of 50% and 100%, compared with reserve and resource data.

Credit: Reserve and resource data from USGS, 2019. USGS mineral commodities survey, US geological surveys. *US-DOI.*

noticeably higher than for NMC 111 and would have consumed 60% of the known reserves (89 million tons) and 40% of the known resources. However, the cumulative demand for Li and Mn by 2050 is slightly lower than that of NMC 111.

Scenario for NMC 811

Fig. 5.5 indicates the cumulative demand for Li, Co, Ni, and Mn for the NMC 822-type battery by 2050. For this battery chemistry, the calculated demand roughly matches the available reserves for cobalt (7 million tons) if recycling (100%) is taken into consideration. However, the Ni demand is noticeably higher in comparison to NMC111 and NMC 622. By 2050, this battery type would have consumed more than 60 million tons of Ni, which is approximately two-thirds of the known reserves, or half of the known resources. Yet the cumulative demand for Li and Mn would be slightly lower than that of NMC 622.

Scenario for NCA

Fig. 5.6 shows the cumulative demand for Li, Co, and Ni for the manufacture of the NCA-type battery. Like NMC 811, NCA represents a high Ni system. Therefore, the Ni demand is again significantly higher than it is for NMC111 and NMC 622 and would involve the consumption of more than 60 million tons, which is slightly

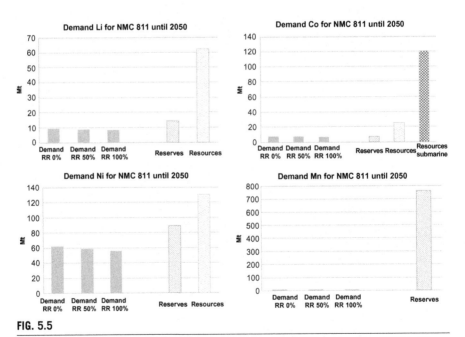

FIG. 5.5

Calculated cumulative Li, Co, Ni, and Mn demand for NMC 811 systems by 2050 in Mt with theoretical recovery rates of 50% and 100%, compared with reserve and resource data.

Credit: Reserve and resource data from USGS, 2019. USGS mineral commodities survey, US geological surveys. *US-DOI.*

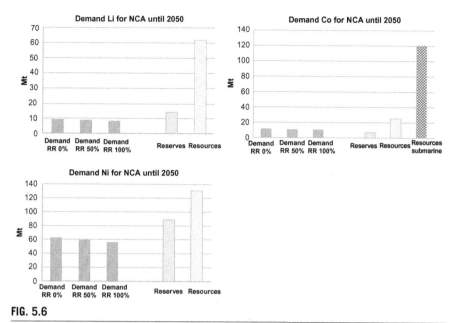

FIG. 5.6

Calculated cumulative Li, Co, and Ni demand for NCA systems by 2050 in Mt with theoretical recovery rates of 50% and 100%, compared with reserve and resource data.

Credit: Reserve and resource data from USGS, 2019. USGS mineral commodities survey, US geological surveys. *US-DOI.*

higher than the amount consumed by NCM 811 batteries. The Co demand is also higher here than it is for NCM 811, reaching 12 million tons or roughly twice the known reserves. The calculated Li demand is comparable to the NMC 811 systems: it amounts to 9 million tons without recycling and 8 million tons with recycling (100%).

Scenario for LFP

Iron and phosphate can be considered abundant resources (Weil et al., 2018; Ziemann et al., 2018). Therefore, in the case of the LFP-type battery, we have only highlighted the cumulative demand for Li and Ni. LFP batteries have the highest total demand for Li among all the battery systems under investigation. This is basically due to their lower energy density, which necessitates larger battery sizes and thus more resources for the same storage capacity. Total Li demand is approximately 28 million tons, i.e., twice the known reserves, or 45% of the known resources (Fig. 5.7). In contrast, the Ni demand is very low and amounts to only 1 million tons. (Ni is not a constituent of the battery chemistry itself, but is required for auxiliary parts such as connectors.)

Discussion of the scenarios

The scenarios presented here do not claim to provide a realistic estimate of the future resource demand caused by a worldwide energy transition. Instead, they are designed to help us understand the resource intensity of specific battery types for stationary applications.

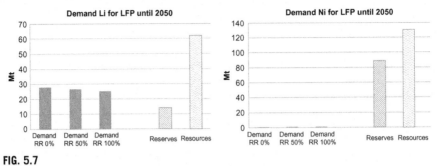

FIG. 5.7

Calculated cumulative Li and Ni demand for LFP systems by 2050 in Mt with theoretical recovery rates of 50% and 100%, compared with reserve and resource data.

Credit: Reserve and resource data from USGS, 2019. USGS mineral commodities survey, US geological surveys. *US-DOI.*

The three types of NMC batteries show a decrease in the required Co content (Table 5.1). This trend has a significant positive effect on the cumulative Co demand by 2050, causing it to decrease from 32 million tons (29 million tons with 100% recycling) to 8 million tons (seven million with 100% recycling). Unfortunately, this comes at the expense of higher Ni requirements. This is also true for the NCA system, which has a low Co content, but high Ni content. Ni is not yet part of the core group of highly critical metals, but its availability is certainly very restricted and it is already a candidate for the EU critical metal list. This became even more obvious after the Indonesian government announced it would stop exporting nickel as a raw material in 2020 (Financial Times, 2019; Jakarta Post, 2019), and the price increased significantly. Thus, the replacement of Co batteries with batteries that require a lot of Ni will reduce, but not solve, the problems associated with the use of restricted resources. In contrast, the LFP system does not need any Co and only small amounts of Ni. In this respect, LFP batteries, which are highly suitable for stationary applications due to their technical performance, seem to be very promising. Unfortunately, LFP batteries cause the highest Li demand among all the evaluated systems: 28 million tons, which is twice the amount of the known reserves. In addition to its restricted availability, the production of Li is also connected with significant environmental and social impacts (as are the other metals in this analysis). A major issue is the high demand for groundwater (in case of brine deposit) in arid regions like Chile, Bolivia, and Argentina. This can create significant problems for agriculture (and society), wildlife, and ecosystems by causing water shortages and/or declines in groundwater levels (Agusdinata, Liu, Eakin, & Romero, 2018). The demand for primary raw materials can be partially reduced by the use of secondary raw materials generated through recycling. However, the results clearly show that even with an extreme theoretical value of 100% recycling (including the whole recycling chain), the recycling effect by 2050 is rather low. One of the key reasons for this is the (estimated) long service life of the batteries. Thus, these results suggest that it is worth researching new-energy-storage systems that are based on more readily available raw materials, which can also be easily recycled.

Recycling and sustainability aspects of LIBs and PLBs

Existing battery technologies for stationary storage vary in terms of their economic aspects, resource demands, and technical performance parameters (cycle life, calendric lifetime, efficiency, and energy density). They also exhibit some major differences in relation to recyclability. These differences are often disregarded when battery technologies are evaluated and compared but can be highly relevant in relation to environmental competitiveness. Recent studies found that a battery system's environmental performance can change significantly if a true life cycle perspective is taken, and the materials potentially recovered by proper recycling are factored in. High recyclability can substantially mitigate disadvantages in other performance categories (such as efficiency or energy density) and, depending on the final application, even lead to better overall performance due to the reduced material demand (Peters & Weil, 2017; Weber, Peters, Baumann, & Weil, 2018).

Recycling of LIBs

Recycling can be considered a key for a future battery industry with minimal material demands and environmental impacts. The main challenges that arise with the existing technologies in this regard are worth mentioning, and also make it possible to chart a course for future developments that take the idea of "design for recyclability" into consideration (see Goddin within this book).

While the production of lead-acid batteries (PbA) is a largely circular industry, the recycling of LIBs is still at an early stage (Mohr et al., 2020). Major large-scale recycling facilities use a rather simple pyrometallurgical process that recovers only a fraction of the materials (the ones with a high market value like cobalt, nickel, and copper) (Mohr et al., 2020). Recycling companies and start-ups are trying to develop and install more sophisticated processes based on hydrometallurgical treatments that also make it possible to recover lower value materials like lithium, graphite, and aluminum (Gaines, 2014). However, these processes increase costs and may cause an economic trade-off situation: greater recycling depth leads to higher costs, while the value from the additionally recovered materials is rather low. In this sense, increasing the recovery rate from 95% to 98%, for example, would require additional process inputs that far exceed the economic (and possibly environmental) benefit. A 100% circular battery economy would, therefore, be neither economically nor environmentally advisable. Instead, an ideal recycling depth could be determined for every battery type that maximizes the environmental (and economic) benefits (Peters, Baumann, & Weil, 2018). While this would be of great interest, not least for future legislation on recycling and circular economies, a significant amount of research still needs to be done before reliable and quantitative recommendations can be made (see Goddin in this book). Especially in relation to economic aspects, the high volatility of raw material market prices would make it difficult to determine clear optimum values.

Another challenge facing the development of a circular battery economy is the decreasing value of recovered materials caused by progress in the battery sector and optimized battery chemistries. These improved chemistries tend to minimize the cobalt content (Choi & Wang, 2018), and in the future will also cut back on other costly metals such as copper (Peters, Peña Cruz, & Weil, 2019). While making batteries more economical and less expensive, this also reduces the value of recoverable materials and might reduce the optimum recycling depth. A good example is the current LFP-type LIB, which does not contain any high-value metals: Its cathode is made of iron, phosphate, and lithium, and its anode is made of graphite. Recycling companies charge original equipment manufacturers fees to compensate for the economic loss involved in recycling LFP batteries (the processing costs exceed the revenue raised by selling the recovered materials) (Peters et al., 2018). In fact, there is no existing market for the "black mass" reclaimed from LFP batteries (recovered cathode material and anode graphite), and recyclers either store it in bulk or send it to landfill. Stationary batteries might have an advantage over automotive batteries in terms of recyclability, because their battery pack casing is less complex and thus cheaper to dismantle. Actually, the dismantling of the often hermetically sealed battery packs is a major challenge and recycling cost factor. If stationary batteries had less strict safety requirements, they would be significantly more economical to dismantle. However, this only relates to the pack dismantling; the recycling of the cells is identical, so the same trade-offs apply. Fundamental advances could be achieved by new battery technologies that are being investigated and developed (this will be discussed in the next section). New technologies have the potential to overcome one of the fundamental drawbacks of current LIBs: The liquid, fluorine-containing and highly flammable electrolyte that poses severe challenges for handling, storage, and processing (recycling) of the battery cells. When opened, the batteries release hydrogen fluoride (HF), which is highly corrosive and toxic and necessitates expensive corrosion-resistant equipment and complex exhaust-gas cleaning. Finally, stationary batteries are often designed to last for a long time, while maximum energy density is less relevant. However, long service lives also cause higher temporal uncertainty in the recycling process. A stationary battery manufactured and installed today might last for years, and it is unknown whether the materials used for the battery will be of any relevance when they are recycled in 20 years. Cobalt, which is now experiencing peak demand (and prices), might have been totally eliminated from LIBs by then—after all, current R&D efforts aim to minimize the cobalt content in batteries. This could lead to an excess supply of cobalt from recycled batteries in the future, which would cause the price of cobalt (and its economic benefits) to drop significantly.

PLBs and potential recycling challenges

The battery sector is progressing at a rapid pace. A decade ago, LIBs were only relevant for handheld and portable devices. Since then, this battery technology has experienced a stunning learning curve and a corresponding drastic decrease in price, so that it is now the technology of choice for automotive and stationary battery storage systems (IRENA, 2017; Strategen Consulting LLC, 2016). Current developments are aiming to further reduce the costs, minimize the cobalt content, and introduce water-based binders and environmentally friendly electrode synthesis processes (Choi & Wang, 2018; Corcuera, Estornés, & Menictas, 2015; Pistoia, 2014; Wu, Ha, Prakash, Dees, & Lu, 2013). However, intensive basic research is also being conducted to identify and develop disruptive technologies that will overcome some of the conceptual limitations of LIBs. Among these are postlithium batteries (PLB), such as sodium-ion batteries and magnesium batteries (Ellis & Nazar, 2012; Hwang, Myung, & Sun, 2017; Saha et al., 2014; Zhao-Karger et al., 2016), but also new lithium-based technologies like lithium-sulfur and lithium-air batteries (Chen & Shaw, 2014; Christensen et al., 2011; Deng, Li, Li, Gao, & Yuan, 2017; Gallagher et al., 2014), as well as solid-state batteries (Kim et al., 2015; Liang et al., 2019). Although there are high hopes for these new technologies, none have yet proved to have the potential to significantly surpass the existing LIB technology. However, from a circular economy perspective, it makes sense to consider their potential recyclability at a very early stage in the development process and to predict possible implications for the battery sector. Currently, the most promising future battery technologies for stationary applications are solid-state batteries (SSB) and postlithium batteries (PLB). The latter is a broad field and currently, the most promising types are sodium-ion batteries (SIB), magnesium-ion batteries (MIB), and metal-air batteries or aqueous hybrid-ion batteries (AHIB).

The potential implications of these future battery technologies on resource demand and recycling are discussed in the following. Development of these technologies is at an early stage and little is yet known about their implications, which is why they are discussed rather speculatively here.

SSB: Solid-state batteries can be divided into two categories: ceramic and polymeric (Kurzweil & Garche, 2017). Both have the advantage of not containing any liquid electrolyte and being compatible with the use of metal anodes, thus offering a potentially high level of safety and high energy densities (and, consequently, reduced material demand per unit of storage capacity). The main drawback that still needs to be overcome is the low ionic conductivity of the electrolytes and interfaces, which leads to poor power capacities and low efficiencies (Meesala, Jena, Chang, & Liu, 2017). Regarding their end-of-life stage, the solid electrolytes do not require costly safety precautions related to fire hazards during storage and handling. Nor do they need expensive exhaust gas cleaning measures and precautionary measures for dealing with hydrogen fluoride emissions and the corresponding corrosion issues—which is one of the major difficulties involved in recycling common LIBs (Peters et al., 2018). Also, the use of metallic anodes (instead of graphite) reduces the material complexity and thus the number of substances that need to be separated during the recycling process. However, there are concerns about the actual recyclability of some of the materials used to manufacture SSBs. Ceramic materials, in particular, are often doped with transition metals (e.g., gallium or lanthanum) to increase their ionic conductivity (Meesala et al., 2017). These transition metals are comparably scarce and increase the impact of the material demand. Furthermore, recovering them from the ceramic matrix is expected to be extremely difficult and only possible using highly complicated processes (if at all) (Peters et al., 2018). Polymeric SSBs appear to have an advantage over ceramic SSBs here, although it is not known whether the recycling of polymeric electrolytes is feasible at all. In any case, the development of these batteries is still in its infancy, so it is too early to reasonably estimate their recycling potential, and so far no research results have been published in this field.

PLB: The term "postlithium battery" refers to any new battery technology that is designed to overcome the limitations of existing LIBs by substituting lithium for another shuttling ion. PLBs are conceptually similar to LIBs in that they consist of sandwiched solid anodes and cathodes, and a liquid electrolyte in sealed cells. Prominent examples are sodium-ion batteries (SIB) (BCC Research, 2018; Hwang et al., 2017), magnesium-ion batteries (MIB) (Mohtadi & Mizuno, 2014; Saha et al., 2014), and aqueous hybrid-ion batteries (AHIB) (Peters & Weil, 2017; Whitacre et al., 2015). While sometimes hailed as the inevitable future market leader due to their good performance and use of exclusively noncritical and abundant materials (Nathan, 2015), PLBs still need to achieve competitively high energy densities, cycle stability, and cycle efficiency. This is currently not yet the case, which is why, despite requiring no scarce materials, they often have higher resource demands than existing LIBs (because a larger battery size is required to provide the same storage capacity) (Peters, Buchholz, Passerini, & Weil, 2016). In contrast, materials that are shown to have outstanding performance are often doped with rare or scarce metals that counteract their potential advances. For instance, AHIB, an environmentally friendly and easy to recycle battery technology, is still inferior to LIB in terms of resource demand and environmental impacts, because of the high battery mass required to deliver a certain storage capacity (Peters & Weil, 2017). SIB and MIB are known as "drop-in technologies" that are based on the same working principles as existing LIBs (Mohtadi & Mizuno, 2014; Peters et al., 2016), which is why they cannot be expected to achieve any breakthroughs in terms of recyclability (same layout, same electrolyte, same problems). Their metallic anodes might be an advantage, as it means there is no graphite to separate and recycle, but this is unlikely to make a significant difference, or rather, this has to be investigated in the future from a sustainability perspective.

a-LIB: Advanced LIBs are new cell technologies that use lithium as the shuttling ion. These include developments such as lithium-sulfur (Li-S) batteries and lithium-oxygen ($Li-O_2$)/lithium-air (Li-air) batteries. They all use metallic lithium in the anode (a key prerequisite for achieving higher energy densities), which reduces the material complexity (no graphite). On the other hand, they still rely on liquid electrolytes, one of the key challenges facing recycling processes. While it might be possible in the future to move away from fluorinated compounds, there are currently no other electrolytes that come close in terms of performance. Li-S batteries contain very low-value materials (no cobalt, nickel, etc.), so the costs associated with this technology will likely be driven by its complex manufacturing process instead. As a consequence, the economic incentive for recycling is low, and recovery might focus on lithium and the metals from the cell housing, thus reducing recycling rates. Li-air batteries do away with the cathode altogether. Instead, they use oxygen (from air) as a cathode material. Oxygen cathode is sensitive to humidity and other trace substances contained in air, so the batteries require either an air separation unit for providing pure oxygen to the cathode, or a pressurized oxygen storage tank (Gallagher et al., 2014; Grande et al., 2015). Both options significantly add to the total battery mass, which needs to be factored in when considering their material demand and recyclability. The oxygen supply is the main drawback of these batteries: it appears to prevent them from fulfilling the promise of their extraordinarily high energy densities. The recyclability of the air separation units and the pressure tanks will depend on the materials and technologies used (e.g., steel or carbon fiber tank). In summary, recycling processes used for existing LIBs may well be applicable to a-LIBs, while the low value of the resulting recycled products might be an obstacle that will require appropriate legislation. Yet if the extraordinarily high proposed energy densities are ever achieved, a-LIBs will significantly reduce the material demands per unit of storage capacity.

Conclusions

In addition to offering a variety of flexible options for the grid, stationary batteries are a key component for a successful energy transition in the future and for reaching the Paris Agreement's ambitious goals for greenhouse gas reduction. However, the potentially high demand for stationary batteries could cause some resource shortages, even if we exclude the enormous demand for batteries expected in the automotive sector. LIBs seem to be among the most promising technologies for this purpose, at least in the mid-term. However, the various LIB technologies not only differ significantly in relation to their technical performance and costs but also regarding their metal content. The present estimation of the hypothetical, cumulative resource demand by 2050 for the different LIB chemistries reveals potential future resource shortages. Some of these shortages can already be witnessed today. Cobalt is generally considered to be a highly critical resource. Moreover, recent reports in newspapers and magazines often point out the environmental pollution, poor working conditions, and even child labor connected with the mining of cobalt (see Prause in this book). The high cost of cobalt is the main driver behind a trend to reduce or even eliminate the cobalt content in battery cells. An example of this trend is the development of NMC 811 and NCA (Tesla) batteries, which have a low Co content. Nevertheless, the reduction of Co is compensated by, in this case, a noticeably increased use of Ni—another potentially critical metal. This could lead to the limited availability of Ni in the future, particularly if such systems were also used in large quantities for stationary applications. An exception here is the LFP system. It only requires one resource with noticeably restricted availability and a high vulnerability potential: Li (Weil & Ziemann, 2014). But all the evaluated systems that require Li have to deal with the resource's restricted availability (and production capacities). There is also a growing awareness in society regarding the adverse environmental and social impacts of Li production, e.g., in South America, which is having a further negative impact on supply.

Thus, the careful selection of the most suitable technology is vital for avoiding excessive material demands and environmental impacts. It is quite important, especially in the long run, that all the components (including graphite) have high recycling rates, as this will reduce the primary resource demand. But the results also reveal that, even with high recycling rates, the cumulative recycling effects by 2050 will be rather small, because of the strongly increasing demand for batteries coupled with the technologies' long service life.

Emerging energy storage technologies based on more abundant materials like magnesium or sodium seem quite promising. Yet here, too, recycling is an important issue.

In general, batteries must feature a sustainable design that minimizes the material demand per unit of energy stored over the battery's service life and includes a thorough recycling concept (unlike today's LIB industry, where lithium is treated like a single-use resource and often discarded after a battery's end of life). Finally, it should be mentioned that the use of batteries for stationary applications requires a thermal management system, and many electric and electronic components (e.g., battery management system or inverter). These technologies are also based on some critical or potentially critical resources and could cause significant environmental impacts, which will be considered in detail in future studies.

By taking a more comprehensive life cycle perspective, it becomes clear that material efficiency measures alone are not enough—there also needs to be a reduction in the demand for stationary batteries. This can be achieved through better sector coupling or better interconnections, and the use of other flexibility options (e.g., load management, improved energy consumption forecasts for more efficient energy provision). But most importantly, a reduced need for the electricity itself will contribute to an environmentally friendly energy transition without having major impacts on the environment.

Acknowledgment

The authors would like to thank the Helmholtz project "Energy System 2050" for providing financial support.

References

Agusdinata, D. B., Liu, W., Eakin, H., & Romero, H. (2018). Socio-environmental impacts of lithium mineral extraction: towards a research agenda. *Environmental Research Letters*, *13*, 123001. https://doi.org/10.1088/1748-9326/aae9b1.

BCC Research. (2018). *Global market for sodium-ion batteries to nearly triple in value by 2022. BCC Mark. Res. Rep.*https://globenewswire.com/news-release/2018/01/18/1296529/0/en/Global-Market-for-Sodium-ion-Batteries-to-Nearly-Triple-in-Value-by-2022.html. (Accessed November 10, 2018).

Chen, L., & Shaw, L. L. (2014). Recent advances in lithium-sulfur batteries. *Journal of Power Sources*, *267*, 770–783. https://doi.org/10.1016/j.jpowsour.2014.05.111.

Choi, S., & Wang, G. (2018). Advanced lithium-ion batteries for practical applications: Technology, development, and future perspectives. *Advanced Materials Technologies, 3,* 1700376. https://doi.org/10.1002/admt.201700376.

Christensen, J., Albertus, P., Sanchez-Carrera, R. S., Lohmann, T., Kozinsky, B., Liedtke, R., et al. (2011). A critical review of Li/air batteries. *Journal of the Electrochemical Society, 159,* R1–R30. https://doi.org/10.1149/2.086202jes.

CNESA. (2018). *CNESA storage market analysis—2017 Q4.* China Energy Storage Alliance.

Corcuera, S., Estornés, J., & Menictas, C. (2015). *Advances in batteries for medium and large-scale energy storage.* Elsevier.

Deng, Y., Li, J., Li, T., Gao, X., & Yuan, C. (2017). Life cycle assessment of lithium sulfur battery for electric vehicles. *Journal of Power Sources, 343,* 284–295. https://doi.org/10.1016/j.jpowsour.2017.01.036.

Ellis, B. L., & Nazar, L. F. (2012). Sodium and sodium-ion energy storage batteries. *Current Opinion in Solid State and Materials Science, 16,* 168–177. https://doi.org/10.1016/j.cossms.2012.04.002.

Financial Times. (2019). *Nickel prices hit four-year high on Indonesia export ban.* Financial Times.

Gaines, L. (2014). The future of automotive lithium-ion battery recycling: Charting a sustainable course. *Sustainable Materials and Technologies, 1–2,* 2–7. https://doi.org/10.1016/j.susmat.2014.10.001.

Gallagher, K. G., Goebel, S., Greszler, T., Mathias, M., Oelerich, W., Eroglu, D., et al. (2014). Quantifying the promise of lithium–air batteries for electric vehicles. *Energy & Environmental Science, 7,* 1555. https://doi.org/10.1039/c3ee43870h.

Grande, L., Paillard, E., Hassoun, J., Park, J.-B., Lee, Y.-J., Sun, Y.-K., et al. (2015). The lithium/air battery: Still an emerging system or a practical reality? *Advanced Materials, 27,* 784–800. https://doi.org/10.1002/adma.201403064.

Hesse, H. C., Schimpe, M., Kucevic, D., & Jossen, A. (2017). Lithium-ion battery storage for the grid—A review of stationary battery storage system design tailored for applications in modern power grids. *Energies, 10,* 1–42.

Hwang, J.-Y., Myung, S.-T., & Sun, Y.-K. (2017). Sodium-ion batteries: present and future. *Chemical Society Reviews, 46,* 3529–3614. https://doi.org/10.1039/C6CS00776G.

IRENA. (2017). *Electricity storage and renewables: Costs and markets to 2030.* Abu Dhabi: International Renewable Energy Agency.

Jakarta Post. (2019). *Indonesia to ban nickel exports from January 2020.* Jakarta Post.

Kaikkonen, L., Venesjärvi, R., Nygård, H., & Kuikka, S. (2018). Assessing the impacts of seabed mineral extraction in the deep sea and coastal marine environments: Current methods and recommendations for environmental risk assessment. *Marine Pollution Bulletin, 135,* 1183–1197. https://doi.org/10.1016/j.marpolbul.2018.08.055.

Kim, J. G., Son, B., Mukherjee, S., Schuppert, N., Bates, A., Kwon, O., et al. (2015). A review of lithium and non-lithium based solid state batteries. *Journal of Power Sources, 282,* 299–322. https://doi.org/10.1016/j.jpowsour.2015.02.054.

Kurzweil, P., & Garche, J. (2017). 2—Overview of batteries for future automobiles. In J. Garche, E. Karden, P. T. Moseley, & D. A. J. Rand (Eds.), *Lead-acid batteries for future automobiles* (pp. 27–96). Amsterdam: Elsevier. https://doi.org/10.1016/B978-0-444-63700-0.00002-7.

Liang, J., Luo, J., Sun, Q., Yang, X., Li, R., & Sun, X. (2019). Recent progress on solid-state hybrid electrolytes for solid-state lithium batteries. *Energy Storage Materials,* https://doi.org/10.1016/j.ensm.2019.06.021.

Masson-Delmotte, V., Zhai, P., Pörtner, H. O., Roberts, D., Skea, J., Shukla, P., et al. (2019). Summary for Policymakers. In *Global warming of 1.5°C. An IPCC special report on the impacts of global warming of 1.5°C above pre-industrial levels*: IPCC.

Meesala, Y., Jena, A., Chang, H., & Liu, R.-S. (2017). Recent advancements in Li-Ion conductors for all-solid-state li-ion batteries. *ACS Energy Letters, 2*, 2734–2751. https://doi.org/10.1021/acsenergylett.7b00849.

Miao, Y., Hynan, P., von Jouanne, A., & Yokochi, A. (2019). Current Li-Ion battery technologies in electric vehicles and opportunities for advancements. *Energies, 12*, 1074. https://doi.org/10.3390/en12061074.

Mohr, M., Weil, M., Peters, J. F., & Wang, X. (2020). Recycling of lithium ion batteries. In *Encyclopedia of electrochemistry: Batteries* [in press].

Mohtadi, R., & Mizuno, F. (2014). Magnesium batteries: Current state of the art, issues and future perspectives. *Beilstein Journal of Nanotechnology, 5*, 1291–1311. https://doi.org/10.3762/bjnano.5.143.

Nathan, S. (2015). *Sodium-ion batteries "set to challenge" dominant lithium-ion technology.* The Engineer.

Peake, S. (2012). What is a low-carbon society? In H. Herring (Ed.), *Living in a low-carbon society in 2050, energy, climate and the environment series* (pp. 15–27). London, UK: Palgrave Macmillan. https://doi.org/10.1057/9781137264893_2.

Peters, J. F., Baumann, M., & Weil, M. (2018). Recycling aktueller und zukünftiger Batteriespeicher: Technische, ökonomische und ökologische Implikationen: Ergebnisse des Expertenforums am 6. Juni 2018 in Karlsruhe (Workshop report No. 99). In *KIT Scientific Working Papers. Karlsruhe Institute of Technology (KIT), Karlsruhe, Germany.*

Peters, J., Buchholz, D., Passerini, S., & Weil, M. (2016). Life cycle assessment of sodium-ion batteries. *Energy and Environmental Science, 9*, 1744–1751. https://doi.org/10.1039/C6EE00640J.

Peters, J. F., Peña Cruz, A., & Weil, M. (2019). Exploring the economic potential of sodium-ion batteries. *Batteries, 5*, 10. https://doi.org/10.3390/batteries5010010.

Peters, J. F., & Weil, M. (2017). Aqueous hybrid ion batteries—an environmentally friendly alternative for stationary energy storage? *Journal of Power Sources, 364*, 258–265.

Pistoia, G. (Ed.), (2014). *Lithium-ion batteries. Advances and applications.* Amsterdam: Elsevier.https://doi.org/10.1016/B978-0-444-59513-3.01001-3.

Ram, M., Bogdanoy, D., Aghahosseini, A., Gulagi, A., Oyewo, A. S., Child, M., et al. (2019). *Global energy system based on 100% renewable energy—Power, heat, transport and de-salination sectors.* Lappenranta, Berlin: Study by Lappeenranta University of Technology and Energy Watch Group.

Saha, P., Datta, M. K., Velikokhatnyi, O. I., Manivannan, A., Alman, D., & Kumta, P. N. (2014). Rechargeable magnesium battery: Current status and key challenges for the future. *Progress in Materials Science, 66*, 1–86. https://doi.org/10.1016/j.pmatsci.2014.04.001.

Strategen Consulting LLC. (2016). *DOE Global Energy Storage Database.*

Tsiropoulos, I., Tarvydas, D., & Lebedeva, N. (2018). *Li-ion batteries for mobility and station-ary storage applications (EUR—Scientific and Technical Research Reports).* Publications Office of the European Union, Petten, The Netherlands: European Commission, Joint Research Centre.

USGS. (2019). *USGS mineral commodities survey, US geological surveys.* US-DOI.

Vaalma, C., Buchholz, D., Weil, M., & Passerini, S. (2018). A cost and resource analysis of sodium-ion batteries. *Nature Reviews Materials, 3*, 1–11. https://doi.org/10.1038/natrevmats.2018.13.

van Wijk, J. M. (2018). *Breakthrough solutions for mineral extraction and processing in extreme environments, Blue Mining public report.* Delft, NL: Blue Mining.

Weber, S., Peters, J., Baumann, M. J., & Weil, M. (2018). Life cycle assessment of a vanadium-redox-flow battery. *Environmental Science & Technology, 52,* 10864–10873. https://doi.org/10.1021/acs.est.8b02073.

Weil, M., Peters, J. F., & Baumann, M. (2019). Resource demand for the mobility transition—Batteries for bus and autonomous minibus vehicles. In *Presented at the 13th Society and Materials International Conference (SAM13).* Pisa: Italy.

Weil, M., & Tübke, J. (2015). *Energiespeicher für Energiewende und Elektromobilität. Entwicklungen, Herausforderungen und systemische Analysen. (Vol. 24).* (pp. 4–9) Technikfolgenabschätzung—Theorie und Praxis.

Weil, M., & Ziemann, S. (2014). Recycling of traction batteries as a challenge and chance for future lithium availability. In *Advances and Applications* (pp. 509–528). Amsterdam, The Netherlands: Elsevier.

Weil, M., Ziemann, S., & Peters, J. F. (2018). The issue of metal resources in li-ion batteries for electric vehicles. In G. Pistoia & B. Liaw (Eds.), *Behaviour of lithium-ion batteries in electric vehicles* (pp. 59–74). Cham: Springer International Publishing.

Whitacre, J. F., Shanbhag, S., Mohamed, A., Polonsky, A., Carlisle, K., Gulakowski, J., et al. (2015). A polyionic, large-format energy storage device using an aqueous electrolyte and thick-format composite $NaTi_2(PO_4)_3$/activated carbon negative electrodes. *Energy Technology, 3,* 20–31. https://doi.org/10.1002/ente.201402127.

Wu, Q., Ha, S., Prakash, J., Dees, D. W., & Lu, W. (2013). Investigations on high energy lithium-ion batteries with aqueous binder. *Electrochimica Acta, 114,* 1–6. https://doi.org/10.1016/j.electacta.2013.09.068.

Zhao-Karger, Z., Lin, X.-M., Bonatto Minella, C., Wang, D., Diemant, T., Behm, R. J., et al. (2016). Selenium and selenium-sulfur cathode materials for high-energy rechargeable magnesium batteries. *Journal of Power Sources, 323,* 213–219. https://doi.org/10.1016/j.jpowsour.2016.05.034.

Ziemann, S., Müller, D. B., Schebek, L., & Weil, M. (2018). Modeling the potential impact of lithium recycling from EV batteries on lithium demand: A dynamic MFA approach. *Resources, Conservation and Recycling, 133,* 76–85. https://doi.org/10.1016/j.resconrec.2018.01.031.

Zubi, G., Dufo-López, R., Carvalho, M., & Pasaoglu, G. (2018). The lithium-ion battery: State of the art and future perspectives. *Renewable and Sustainable Energy Reviews, 89,* 292–308. https://doi.org/10.1016/j.rser.2018.03.002.

Making critical materials valuable: Decarbonization, investment, and "political risk"

6

Paul Robert Gilbert

International Development, University of Sussex, Brighton, United Kingdom

Introduction

In this chapter, I draw on two strands of work within the field known as science and technology studies (STS) to shed light on the relationship between critical raw materials assessments and patterns of (speculative) extractive industry investment. Firstly, I engage with work on *resource-making* (Ferry & Limbert, 2008; Kama, 2019; Richardson & Weszkalnys, 2014), which traces how forms of expertise, legal arrangements, economic considerations, and wider "resource imaginaries" create the conditions under which substances can exist as resources and thus be drawn into a range of economic relations. Secondly, I draw on the emerging subfield known as *valuation studies*. This body of work rejects attempts to identify the "true" value of objects or relations and instead examines the evaluative infrastructures made up of narratives, rankings, indices, and formulae that *apportion* value as they *measure* it (Muniesa, 2011). In other words, this approach is attentive to the practices that simultaneously *calculate* value and *make things valuable* (Kornberger, Justesen, Madsen, & Mouritsen, 2015). I bring these approaches to resource-making and valuation practices to bear on methodologies for critical raw materials assessment, focusing in particular on the role that measurements of "political risk" play in determining the "criticality" of materials required for ushering in a number of possible decarbonized energy futures.

I focus on the role that political risk metrics play in criticality assessments because these measures (and the very notion of "political risk" itself) reflect the priorities of extractive industry corporations and investors based in global centers of mining finance (notably London and Toronto). The discourse and measurements of political risk present a wide range of actions that might be undertaken by postcolonial states as aberrant precisely *because* they constitute threats to the valuation of mineral deposits and mining assets owned by transnational mining corporations (Gilbert, 2020a, 2020b). The normalization of the language of political risk—which has been disseminated from conservative think-tanks aligned with extractive industry corporations into criticality assessments carried out by energy researchers concerned with decarbonization—also underpins the ongoing historical erasure of attempts that

The Material Basis of Energy Transitions. https://doi.org/10.1016/B978-0-12-819534-5.00006-4

formerly colonized states made to inaugurate a more just international economic order (Gilbert, 2018).

After introducing the literature on resource-making and valuation studies, and bringing this work to bear on the methodology of criticality assessments and their reliance on measures of political risk, this chapter turns to two case studies (one from Democratic Republic of Congo, and one from South Africa) which reveal that ongoing attempts to make critical raw materials valuable continue to reinforce unjust postcolonial forms of resource extraction and extractive industry financing. Equally, these case studies highlight the importance of "provincializing" critical raw materials assessments which almost always take the concerns of powerful resource-*consuming* countries in the Global North (United States, Japan, EU-27) as the location from which to determine criticality. In this context, provincializing critical raw materials assessments means recognizing that thought is located to place and that by starting with the concerns about critical or "strategic" minerals expressed by the DRC or South Africa, we can question the taken-for-granted ideas that currently underpin criticality assessments, broadening their scope, and situating them in relation to the concerns of resource-rich postcolonial countries (Chakrabarty, 2000; Lawhon, Ernstson, & Silver, 2013).

"Resource-making" and critical raw materials

For STS scholars, artifacts such as mobile phones, batteries, and electric vehicles can present themselves as guides to unpacking the materials and relations that make up a global political economy (Dumit, 2014). They invite questions about labor (who produced this, where, under what conditions?), expertise (who makes decisions about or enables innovation in relation to each of its components?), materials (where did they come from, how are they disposed of?) and economy (where are its components sold, by whom, and how are costs and risks calculated?). The ever-increasing complexity of electrical devices is therefore as much sociological as it is technological. More components not only increase their mineral intensity, but multiplies the connections that these devices have to differing labor regimes, economic processes, and forms of expertise and authority. Critical raw materials assessments conducted by the United States and the EU also respond to the growing componential complexity and mineral intensity of emerging technologies (e.g., Hollins & Fraunhofer, 2013, p. 5, see Wang et al. within this book) although they are guided by anxieties over resource security and economic growth rather than a desire to reveal the global political-economic relations that are condensed in everyday objects.

Concerns about reliance on an ever-greater number of imported nonfuel materials prompted the Committee on Earth Resources of the US National Research Council (NRC) to develop the first "criticality matrix," where criticality is a product of economic importance and supply risk (National Research Council, 2008) (see also Wang et al. and Zepf within this book). The EU followed with their review of Critical Raw Materials (CRMs) in 2010 (updated in 2014 and 2017), introducing cutoff levels for supply risk and economic importance that taken together define criticality

(European Commission, 2017a, 2017b). These initial national/regional criticality assessments were not particularly preoccupied with materials necessary for decarbonizing energy production. Subsequent work has however developed criticality assessment methodologies with a specific focus on "clean" energy technologies. This has moved on from the criticality matrix and economy-level measures of importance used by the United States and EC in favor of product design tree approaches that present the criticality of different wind turbine designs (Habib & Wenzel, 2016); life-cycle assessments that examine the criticality of materials required for the production of electric vehicles within the EU (Cimprich et al., 2017); national-level "mining risk footprint" analyses that build on material footprint analyses (Nansai et al., 2015); and dynamic analyses of projected demand for various critical materials based on possible scenarios for decarbonizing energy production (Giurco, Dominish, Florin, Watari, & McLellan, 2019; Ioannidou, Heeren, Sonnemann, & Habert, 2019; Watari, McLellan, Ogata, & Tezuka, 2018) (see also Weil et al. and Penaherrera and Pehlken within this book).

Given that criticality can be calculated for product, supply chain, national, regional or global scales, and using either static measures or dynamic scenarios, it comes as little surprise that a "harmonized and standard definition of criticality does not exist," and that "methodologies for the assessment of CRMs have to implicitly answer the question: critical to whom" (Mancini, Benini, & Sala, 2018, pp. 728–729). And, as Habib and Wenzel (2016, p. 3854) note, a *mineral* is not critical by definition; rather criticality is a property of a (socio-technical) system which creates conditions for criticality. Here we find further parallels between methodologies for criticality assessment and work on "resource-making" in STS, anthropology, and geography. Beginning with the premise that "nothing is essentially or self-evidently a resource" (Ferry & Limbert, 2008, p. 3; see Huber, 2018, p. 553), a growing body of work in this area has turned attention, not to substances themselves, but the "complex arrangements of physical stuff, extractive infrastructures, calculative devices, discourses of the market and development, the nation and the corporation…that allow those substances to exist as resources" (Richardson & Weszkalnys, 2014, p. 7).

In her recent state-of-the-art review of approaches to resource-making controversies, Kama (2019) examines how the "resourceness" of unconventional hydrocarbons (such as shale gas and tar sands) takes shape in response to a temporal horizon that projects their future resourceness as inevitable (a mere product of knowledge, technical innovation, and constrained availability of conventional hydrocarbons) albeit open. Yet the eventual resourceness of unconventional hydrocarbons depends upon geoscientific knowledge controversies, as various interpretations of resource availability and malleability compete. It equally depends upon on wider political-economic controversies, as rival attempts are made to render unconventionals valuable, or undermine the valuation of these resources by pointing to the likely environmental consequences of their extraction. Certain parallels can be drawn between resource-making around unconventional hydrocarbons and critical raw materials assessments, particularly regarding the role of expertise, and the imagined futures upon which "resourceness" depends.

Criticality assessments have been published by government agencies in order to encourage public and private sector actors to take appropriate steps to "mitigate potential restrictions in the supply of that mineral for an existing, or future, use" (National Research Council, 2008, p. 3). As Kama (2019) notes for unconventional hydrocarbons, expectations about (potentially inevitable) future use condition the circumstances under which substances become resources, as well as the basis on which they are valued (see below). More broadly, "resource imaginaries" imbued with hope, dread or nostalgia play a crucial role in resource-making, framing the present and future in certain such that substances are apprehended as (critical) resources (Ferry & Limbert, 2008, p. 4). Energy demand projections and renewable energy scenarios play a significant role as resource imaginaries underpinning criticality assessments (see McLellan and Weil et al. within this book). For example, several authors have referred to the International Energy Agency's (2010) BlueMAP scenario for increased wind energy use and fuel cell vehicles by 2050 as the basis for assessing criticality (Habib & Wenzel, 2016; Ioannidou et al., 2019), even though the IEA does not itself include assessments of critical raw material demand in its projections (Watari et al., 2018). It becomes apparent that future energy scenarios like those provided by the IEA are profoundly political "resource imaginaries" when considering that they do not predict meeting the 2015 Paris targets of limiting temperature increases to 1.5°C over preindustrial levels by 2050. As Giurco et al. (2019, pp. 445–446) show, assuming the Paris targets are met and based on a goal of 100% renewable electricity and transport, projected demand for cobalt and lithium will exceed current economically viable resources *and* total known reserves by 2022–23.

In fact, IEA projections have been shown to consistently underestimate the actual growth rates in solar PV and wind energy in the projections they published between 1994 and 2014 (Metayer, Breyer, & Fell, 2015). Their 2018 World Energy Outlook, which is routinely used to shape investment demand and policies, has set a course to exhausting the 1.5°C carbon budget by 2022 (Oil Change International & IEEFA, 2018). Work in STS and related disciplines encourages us to treat energy scenarios like this not as mere models, but as technologies of imagination that help to generate and stabilize expectations about the future, enabling speculative investment in resources whose value depends on the future that investors feel most likely to materialize (Gilbert, 2020b; see also De Ville & Siles-Brugge, 2015). It is apparent then, that clean energy-focused critical materials assessment relies upon energy scenarios or resource imaginaries, in order to mobilize the "resourceness" of materials like lithium, cobalt, and platinum group metals. For the junior mining firms hoping to transform speculative lithium, cobalt, or platinum deposits into profitable mines, projections of future critical raw material demand "operate as 'gestures' at future prospectivity, which alongside reserve inventories are not just indispensable in securing the firm's value and its liquidity but help to induce the resource out of mere geologic propensities and economic conjectures" (Kama, 2019, p. 17). In other words, the proliferation of critical raw materials discourse and assessments has come to play an instrumental role in the processes that make deposits into valuable resources.

Valuation, critical raw materials, and "political risk"

The process of resource-making is inevitably tied up with the process of valuation. This process is often speculative (Kama, 2019; Tsing, 2000), since significant initial investment is required to conduct the surveys and assays through which possible deposits become *capitalized* as mineral assets—that is to say, valued in terms of the future revenues they are likely to produce (Gilbert, 2020a). For Fabian Muniesa and colleagues (Muniesa et al., 2017) capitalization is an operation essential to the realm of finance (typified, for instance, in discounted cash flow models) but is also more than that. Capitalization is, they argue, a widespread "cultural syndrome" according to which "the viability of all things resides, primarily, in the asset condition" (Muniesa et al., 2017, p. 23, 52). That is, things become valuable in light of their capacity to produce a future stream of revenue for their owners. More than this though, Muniesa and colleagues highlight the extent to which valuing things in terms of their future revenue potential *transforms* reality. In relation to the extractive industries, the desire to maximize measures of capitalization that are of paramount interest to potential investors (e.g., net present value, or the sum of discounted revenues that a mineral asset is expected to produce over its lifetime) can prompt extractive industry firms to opt for more flexible labor arrangements in their mine plans, or extract resources more rapidly than host governments might wish them to (Gilbert, 2020a).

A key consideration when extractive industry investors calculate the value (and thus potential "resourceness") of a prospect is the discount rate, which reflects the confidence they have in all future revenue streams actually materializing. If confidence is low, the discount rate will be high, and the net present value (sum of discounted future revenue streams) will be lower, reflecting the notion that owning a risky asset is worth less than owning a safe one. A significant factor influencing the discount rate that is adopted for extractive industry projects is the assessment of "political risk" that inheres in a given jurisdiction, and a key metric that is used to aid this process of discounting is the Fraser Institute's Policy Potential Index (Gilbert, 2020b). The Policy Potential Index (PPI) presents itself as a "report card to governments" on the attractiveness of their jurisdiction, compiled by surveying a pool of junior mining executives. To assemble the PPI, managers and consultants are asked to rate jurisdictions with which they are familiar in terms of 15 criteria, including uncertainty over what will be designated protected areas, uncertainty concerning environmental regulations, legal process, political stability, and taxation. These issues are rated on a scale from "Encourages exploration investment" to "Would not pursue exploration investment in this region due to this factor" before the ratings are transmuted into hierarchical rankings.

Notably, the first critical raw materials assessment carried out by the United States utilized the Fraser Institute's PPI as the basis for calculating political risk or "the likelihood of political decisions that alter the government rules under which mining occurs," flagging, in particular, the risk of "resource nationalism" (National Research Council, 2008, p. 56). Notably, the Fraser Institute presented by the NRC has an "independent research and education organization" (National Research Council, 2008).

Critics, however, have described the Fraser Institute as a market libertarian Canadian think tank noting that, along with the PPI, it created the "neoliberal" Economic Freedom in the World index in collaboration with Milton Friedman (Gutstein, 2014). It is not only the NRC that used the Fraser Institute's PPI rankings to incorporate a notion of political risk into their criticality assessments, but Nansai et al. (2015) also used the PPI in order to generate their measures of "mining risk footprint" for critical raw materials essential for renewable energy production in Japan.

Rather than treating these metrics as "true" measures of the "political risk" that characterizes particular jurisdictions, rankings like these that are produced by surveys of executives and consultants can be understood as attempts to generate "resource imaginaries" according to which environments that pose any restrictions on the capitalization of mineral assets are seen as deviant. The explicit disciplinary intent behind the PPI is captured in its self-presentation as a "report card to governments," and a poor ranking seeks to undermine resource-making efforts, putting pressure on host governments to create more generous regulatory environments. Other criticality assessment methodologies follow the EU's approach (Hollins & Fraunhofer, 2013), which uses the World Bank's World Governance Indicators (WGIs), especially the political stability and absence of violence/terrorism metric, to generate a measure of vulnerability to supply risk (e.g., Cimprich et al., 2017; Gemechu, Helbig, Sonnemann, Thorenz, & Thuma, 2017; Habib & Wenzel, 2016). Yet the WGIs are no more "neutral" than the Fraser Institute's PPI. They draw on a range of other indices, some of which also rely on surveys of executives' perceptions of given jurisdictions and thus amplify various prejudices about corruption (Cobham, Janský, & Meinzer, 2015). Other indices on which the WGIs draw have been shown to reflect pro-US Foreign Policy biases, as well as distinct ideological commitments regarding the privileging of civil rights over socioeconomic rights (Bush, 2017).

The incorporation of these metrics into criticality assessments reflects the further dissemination and normalization of a resource imaginary that privileges the concerns of extractive industry firms and investors who seek to maximize the value of their assets and present a host of regulatory acts as hostile, unwarranted or deviant behavior. The notion of "political risk" itself emerged in response to the anxieties of United States and European investors who were forced to deal with newly independent, postcolonial states in the mid-20th century (Kobrin, 1982). As conservative, pro-empire economic historians note, prior to independence the only "political risk" that investors had to face was "the expected duration of [imperial] rule" (Ferguson & Schularick, 2006). Expressions of sovereignty were recast as "risk" factors, especially in relation to the extractive industries, where postcolonial states sought to renegotiate unfavorable concessions and royalty agreements negotiated with preindependence authorities, or nationalize mines in order to save developmental agendas.

Contemporary political risk analysis also relies heavily on the language of political risk *insurance*, which emerged in the City of London in the 1970s, and continues to provide cover for the risks of confiscation, expropriation, and nationalization (CEN)—risk categories that reflect widespread extractive industry anxieties about "resource nationalism" (Gilbert, 2020a). However, successive rulings by international arbitration panels dealing with disputes between extractive industry investors and postcolonial host states have resulted in an expansive definition of "expropriation," such that any government action that impinges upon investors *expectations* might be treated as an act of expropriation. A particularly troubling area for critical scholars of international investment law is the emergent concept of "creeping regulatory expropriation" whereby *no single act* constitutes expropriation, but apparently "expropriatory" taxation rates *may* do so (Kantor, 2015, p. 179). That is, high royalty and taxation rates (as well as stringent environmental regulation) might be treated as grounds for investors to bring legal proceedings against host states, and in some cases, their grievances have been upheld.

By incorporating measures of political risk derived from the Fraser Institute's PPI and the World Bank's WGIs, critical raw material assessments help to normalize and disseminate a resource imaginary according to which governmental action that reduces the prospective value of a mineral resource—or indeed investors' *expectations* that such value might be realized—is deviant and unwelcome. Reproducing metrics of political risk in this way is a byproduct of criticality assessments' preoccupation with the needs of wealthy countries in the Global North (predominantly the United States, the EU, and Japan). The remainder of this chapter, therefore, moves to examine assessments of critical or strategic materials made by countries that *produce* minerals crucial to decarbonizing energy production (DRC and South Africa). The purpose of examining these case studies (Lithium and cobalt in DRC and platinum group metals in South Africa) is twofold: firstly, it highlights the extent to which current models of criticality assessment need to be "provincialized" and decoupled from their commitment to the concerns of the Global North; and secondly, it highlights how current extraction of "critical" materials crucial to decarbonized energy futures is embedded in exploitative postcolonial relationships.

"Strategic" minerals, royalties, and stabilization provisions in DRC

The historical concentration of global cobalt production in DRC is frequently referenced by those attempting to incorporate measures of "political risk" into criticality assessments. Habib et al. (2016, p. 851) note, for instance, that "political instability" in DRC (then Zaire) in 1978 caused a cobalt production slowdown, triggered speculation that led to a fivefold price increase, and incentivized research into the substitution of cobalt in permanent magnets. Cobalt has been categorized as "Critical" in the EU criticality assessments for 2010, 2014, and 2017, based on its importance as a battery chemical and EU reliance on imports (46%) from DRC (European Commission, 2017a, 2017b). The use of cobalt in the cathodes of around 75% of lithium-ion batteries means that any significant decarbonization of transport infrastructure and the rise of electric vehicles (EVs) will only increase demand for cobalt (Giurco et al., 2019) (see Weil et al. within this book). Metals industry analysts note that in the absence of innovations that reduce the cobalt content of EV batteries, there "will be no EV industry without DRC cobalt" (Sanderson, 2019). It is perhaps no surprise, therefore, that DRC designated cobalt a "strategic" mineral under the terms of its new mining code in 2018.

Revisions of DRC's postwar 2002 mining code began in 2012 and resulted in a bill approved by both houses of the legislature in January 2018. Law 18/001 of March 9, 2018, signed into force by the then president Joseph Kabila, amended the 2002 Mining Code. The new 2018 Mining Code has been fiercely resisted by representatives of transnational mining corporations operating in the DRC. Some have threatened arbitration over reductions in the "stability" provisions that prevent changes to tax and royalty rates for new miners (Mahamba & Cocks, 2018), and others like Glencore's Ivan Glasenberg have hinted at legal action regarding changes to royalty rates, stating that "we don't accept them—even the current royalty we're paying under duress" (Mining Journal, 2018). The DRC's secretary-general of mining spoke to the Mining Indaba in Cape Town (one of the industry's three biggest annual events), affirming that the law must be respected and would not be renegotiated, although Barrick Gold CEO Mark Bristow rejected this request and insisted he would continue his conversation on "what is good for the Congo" (Hume, 2019; see Randgold Resources, 2018).

The specific changes that have caused a backlash and threats of arbitration from DRC-based miners include the reduction of a stability period during which fiscal terms cannot be changed from 10 to 5 years (Article 276 revised, new Article 342), although mining journalists have frequently reported simply that the stability clause was "removed" (Jamasmie, 2018). In addition, the new mining code increased royalties for nonferrous metals from 2% to 3.5%, payable regardless of a company's profitability, and with the potential for royalties to be raised to 10% on "strategic" materials designated "on the basis of the Government's opinion of the prevailing economic environment, are of special interest given the critical nature of such mineral and the geo-strategical context" (Articles 241 and 255). As such, on November 24, 2018, cobalt, coltan, and germanium were declared "strategic mineral resources" by Prime Minister Bruno Tshibala because of their high-tech applications and the "current international economic conjuncture" (Clowes, 2018).

It is worth considering the decision made by DRC to designate cobalt "strategic" alongside the EU's designation of cobalt as a "critical" raw material, as well as the response from extractive industry companies, investors, and lawyers, in light of the "political risk" imaginary outlined earlier in this chapter. For instance, the London-based legal firm Norton Rose Fulbright wrote in a February 2018 briefing on the new DRC mining code: "Over the past few years, several African countries have substantially overhauled their mining laws, although some have worked harder at affording *stability protection* to those miners that had believed in their country (often during difficult times) and massively invested to develop world-class mining projects" (Norton Rose Fulbright, 2018, emphasis added). This framing, which presents "stability protection" as a *right* to be afforded to investors, reflects the "neoliberal conception of risk" that Jody Emel and Matthew Huber have shown to be the driving force of World Bank-backed mining code reform in Africa during the 1990s: under this concept of risk, it is made clear that transnational mining corporations and their investors *bear the burden* of risk, and so it is the host states' obligation to provide them with favorable terms and investment incentives (Emel & Huber, 2008, p. 1396).

There is a longer history to this, however, than the "neoliberal" context that Emel and Huber make their focus. Critical scholars of international investment law have argued that the current system of international investment law upholds the legacy of a colonial world order whereby the sovereignty of a "Third World" state was *conditional* on a Euro-American corporation's right to trade, and placed stabilization clauses at the center of this continuing legacy. As Anthony Anghie shows when states and corporations enter into a contract bearing a stabilization clause, this "transform[s] the state into a private actor that was merely contracting with another private actor, and could not rely on any residual sovereign powers to amend the terms of the contract, *whatever the demands of public welfare.*" (Anghie, 2007, p. 234, emphasis added). Violation of such clauses will usually trigger arbitration under a Bilateral Investment Treaty (BIT) between the host state and the state in which the corporation is domiciled (as per the threats made by mining executives cited above), which can be profoundly costly while also undermining judicial authority in the host state by allowing a choice of law from other jurisdictions. Under this system, "the state is pressed to avoid measures that are needed to prevent poverty and encourage sustainable social development through distributive methods of taxation, environmental measures, and observance of human rights standards where those might impact on the foreign investors' profits" (Linarelli, Salomon, & Sornarajah, 2018, p. 161). Specifically, the argument that postcolonial states should maintain the sovereign right to redefine taxation and royalty terms in the context of "windfall profits" or changes in a resource's price, and that stabilization clauses should not impede on their ability to do so, was at the center of the "Third World Approach to International Law," and the ultimately abortive attempt to craft a New International Economic Order through the nonaligned movement at the UN (see Hossain, 1983).

The point I aim to make here is that reactions to DRC's designation of cobalt as a strategic mineral resource betray the persistent coloniality of extractive industry investment and the legal support that is provided to it. Investors' demand for stability, and the attempts they make to circumvent the sovereign authority of DRC who are seeking to secure developmental resources from the extraction of a mineral that is essential to high-tech applications including decarbonized transport/EVs, reflects not only the "neoliberal" notion of risk according to which only investors and extractive industry corporations *bear* the risks of extraction; it relies on a system of international law established during the age of formal European colonialism that acts to circumvent the sovereign authority of resource-rich states, treating them as private contracting parties and transferring sovereign authority to extractive industry corporations domiciled in the global North. As long as criticality assessments focusing on the needs of wealthy Northern countries (the United States, EU, and Japan) incorporate political risk assessments (such as the PPI or WGIs) into their methodologies, they are also incorporating into those assessments a colonial resource imaginary according to which expressions of sovereignty by postcolonial states constitutes a threat to "stability" and supply.

Lithium, unlike cobalt, has *not* been designated a strategic mineral by DRC, despite early speculation that it could have received the same treatment. Lithium is also not designated critical by the EU, based on an economic importance measure that falls below the cutoff on the EU's criticality matrix (European Commission, 2017b). However, as Cimprich et al. (2017, pp. 760–761) note, when considering decarbonization and energy futures, this perhaps reflects a weakness in the EU's economy-level assessment methodology. Lithium is particularly crucial for "clean" energy and EVs even if batteries currently comprise 1% of total EU uses. The EU's failure to categorize lithium as critical also reflects a weakness of *static* criticality assessments which underestimate future demand (see also Overland, 2019). Like cobalt, lithium demand is set to outstrip current resources and reserves if the transport is to be based on 100% renewables by 2050 (Giurco et al., 2019). Because it is not classified as a strategic mineral in the DRC (and this is perhaps linked to the failure to categorize lithium as critical in the EU and other assessments), investors in lithium prospects in the EU have been able to make their resource (appear) valuable to investors by generating large net present value figures, unencumbered by the larger 10% strategic royalty payments. Australian-based AVZ Minerals currently holds exploration permits for the Manono lithium project in DRC, and as a result its drilling campaigns has upgraded the resource to 269 million tons of 1.65% lithium oxide (up from 189.9 million tons) in May 2019, with a total of 400 million tons if "inferred" resources are included (Australian Mining, 2019).

AVZ Minerals has entered into innovative financing arrangements such as "off-take" agreements with Beijing National Battery Technology and Guangzhou Tinci Materials Technology, and has received backing from large financial institutions including Citicorp, JP Morgan, and BNP Paribas. AVZ's scoping study, published in May 2019, reported a net present value of US$2.63 billion, based on relatively depressed lithium prices of US$750 per ton. This scoping study has been celebrated as providing "industry leading" returns, even in the context of bearish forecasts about lithium prices from Morgan Stanley and others (Sanderson, 2018). (Lithium prices have been particularly volatile, and the London Metals Exchange has only recently attempted to establish a global reference price for lithium contract trading.) The interest of large investment banks in Manono, as well as the offtake agreements signed with Chinese battery producers, reflects the importance of the Manono deposit to global lithium supply—and if any serious efforts are made to meet the Paris 1.5°C scenario by 2050, the value of the deposit will only increase as EV requirements outstrip current resources and reserves (Giurco et al., 2019).

AVZ has been able to raise funds and keep momentum behind the Manono project by generating attractive valuations which in turn contribute to the "resource-making" process. In Kama's terms, these valuations gesture toward future resourcefulness, helping to bring forth the resource from geologic potential and economic valuations (Kama, 2019). Perhaps, had lithium been designated as a strategic mineral resource by DRC, and royalties increased from 3.5% to 10%, these valuations would have been less attractive, and certainly might not have been "industry leading." Yet, if we are to take seriously the notion that wealthy countries in the global North are undertaking efforts to decarbonize transport and energy production in order to meet the 1.5°C Paris targets, then we *know* that demand for lithium will increase, as would valuations and the potential for resource-making even in the context of higher royalties. As such, it becomes necessary to ask whether we need to "provincialize" criticality assessments, rooting them in the economic aspirations and concerns of resource-rich postcolonial states as much as in the desire that resource-importing Northern states might have to ensure a "stable" supply.

HySA and South Africa's platinum futures

The second and final case study in this chapter continues the focus on tensions between assessments of critical or strategic minerals in the global North/postcolonial South. It also highlights further the degree to which the extraction of critical raw materials essential for decarbonized energy futures continues to be embedded in colonially derived forms of investment that subordinate the concerns of postcolonial nations and communities to the resource-making and valuation concerns of extractive industry corporations based in the global North. Specifically, this section examines platinum group metals (PGMs) in South Africa, and the tensions between criticality assessments made in the resource-importing global North, and South Africa's agenda for enhancing the beneficiation of strategic minerals domestically. It also examines the way in which the 2012 Marikana Massacre and its possible impact on platinum mining and platinum prices, has been read through the lens of "political risk."

Of relevance to the above discussions of Bilateral Investment Treaties, arbitration provisions, and assessments of "political risk," South Africa took the bold step of unilaterally canceling 23 of its BITs, beginning in 2012, following a 3-year review. The South African government decided to replace these BITs with domestically crafted legislation, responding to a fear that public interest regulation (including postapartheid Black Economic Empowerment agendas and requirements to transfer mining company shares to historically disadvantaged South Africans) was seen as potentially violating BIT requirements (Mossallam, 2015). BITs typically require that foreign investors are treated under "Fair and Equitable Treatment" and "Most Favored Nation" clauses, which in this case prevent regulation designed to redress historical wrongs from being enacted. In the terms introduced above from Emel and Huber (2008), Anghie (2007), and Linarelli et al. (2018), BITs reflect a system of international investment law that requires host governments to rescind elements of their sovereignty and act as private contracting parties, treating mining companies more favorably on the basis that they are deemed to bear the majority of the risk associated with extraction. As critics of this move note, it "does not help the investor one bit" (Schlemmer, 2015), but it does create more space against which the inequities of apartheid, in which major platinum mining companies were complicit and direct beneficiaries (Rajak, 2014) can be addressed.

The inclusion of PGMs, first collectively and then individually, on the EU's list of critical raw materials, is a product of the overwhelming concentration of resources and production in South Africa (European Commission, 2017a). The PGM market is, however, in constant flux, as concentrated supply and production in South Africa articulates with global fabrication capacity, and the desire of fabrication firms to develop "thrift" strategies that reduce platinum concentration in key industrial outputs, thus placing them in a strategic position to determine prices (Ritchken, 2018). Recently, the shift toward electric cars appears to have suppressed platinum demand in the context of already existing oversupply. This appears to be a product of reduced demand for diesel engines containing PGM-based autocatalytic converters, while the projected growth of platinum-rich hydrogen fuel-cell-based EVs has not fulfilled its promise, particularly in China, and some manufacturers have pushed for platinum-light EV fuel cells (Hobson, 2019; Hu, 2019; Onstad, 2019). If the Paris 1.5°C targets are to be met by 2050, however, the uptake of fuel cells in EVs and electricity storage should theoretically enhance demand for PGMs (Giurco et al., 2019).

Precisely what this might mean for South Africa is unclear. While there is significant foreign interest in investing in PGM mining, concerns about the relation between PGM mining and South Africa's broader "sustainable development" are domestic. Ritchken (2018) argues the importance of South Africa positioning itself as a world leader in green technology, catching the wave that might be offered by an uptake in EV fuel cell technology and building on the existing "HySA" strategy as well as planned platinum export development zones. The biggest challenge South Africa faces here is the concentration of PGM-related technological developments outside the country (in Europe and Asia). Hence the HySA (Hydrogen South Africa) strategy that was introduced in 2007, to focus on the "use and displacement of strategic minerals" (Pollet et al., 2014, p. 3578), harnessing South Africa's concentrated supply of PGMs to renewable energy development, with an initial aim of supplying 25% of the global hydrogen infrastructure and fuel cell market by 2020 (Bessarabov et al., 2012).

Once again, as with the decision to classify cobalt as "strategic" on the part of DRC, South Africa's concern with the beneficiation and disposition of "strategic" PGMs shares with EU and other criticality assessments an acknowledgment of the current (and potential future) importance of cobalt/PGMs in the context of decarbonized energy (and especially transport) supply. However, as in the DRC case, there is a need to "provincialize" criticality assessment when it comes to PGMs, and consider what measures of criticality would look like if they *started* in resource-rich countries facing significant development challenges (like South Africa), rather than with the import-reliance of wealthy Northern countries and trading blocs. Noting that the EU raw materials initiative responded to lobbying from European corporations (including those representing significant consumers of South African PGMs, such as BASF who has been a strategic buyer from Lonmin) concerned about intensifying competition from "emerging" (see Gilbert, 2019) economies, Boniface Mabanza asks: "What would actually happen if African states were to develop resistance to trade agreements to such an extent that Europe's supply would actually be at risk?" (Mabanza, 2018, p. 364). Here Mabanza is referencing the EU/Southern African Development

Cooperation (SADC) Economic Partnership Agreement, which demanded significant export duty removals; this both removes the support for developing processing capacity locally (such as through the HySA/PGM cluster) and deprives countries of taxation revenue which could be used for investments in health and education.

Provincializing criticality assessments would require that such considerations were taken into account, with a view to ensuring that the disposition of minerals for "clean" energy technologies did not merely reproduce colonial approaches to securing resource access from colonial subjects "reduced to [their] labor" (Mabanza, 2018, p. 335). Indeed, labor relations become crucial components of "political risk" measures used in criticality assessments, with the EU assessment citing "labor disputes over wages and working conditions" as one factor of PGM supply risk (European Commission, 2017a, p. 249). Metal market analysts also reduce strike action to a driver of supply risk, and a factor necessitating higher metal prices if miners are to proceed with new investments (Whitehouse, 2019), much as they did following the Marikana Massacre in 2012 (Sainsbury, 2012). The Massacre, on August 16, 2012, resulted in the deaths of 44 mineworkers at Lonmin's Marikana platinum mine was (and is) typically discussed in the mining industry press as the result of internecine union conflicts. This is despite evidence that Lonmin had failed to adhere to its legal obligation to provide decent housing for its workers, the fact that Lonmin officers shot at striking workers with rubber bullets while failing to account for this, presented the strike as having taken them by surprise when it emerged from weeks of negotiation and subsequently refused to negotiate with strikers in the run-up to the Massacre (Styve, 2019, pp. 44–52). Further, as Forslund (2015) has shown, ending Lonmin's profit shifting to Bermuda could have allowed them to meet a significant portion of the striking mineworkers' wage demands, while the entire workforce of 4252 rock drillers' wage package was almost equivalent to the pay of Lonmin's top three executives (Styve, 2019, p. 68).

Translating the Marikana Massacre and subsequent strike action into the language of "political risk," and treating it as an impediment to secure raw material supply for wealthy Northern countries, reflects a "resource imaginary" according to which resources are most effectively made valuable when exploitation and violence toward black workers are allowed to persist (Styve, 2019). As Bond (2013, p. 572) has shown, this is precisely what leading credit agencies like Moody's did following the Marikana Massacre, where they announced that any further strikes would be "credit-negative for rated miners with exposure to South Africa." There is neither room within this language of political risk for consideration of Lonmin's failures to meet its legal obligations surrounding decent housing, nor of the economic partnership agreements that privilege EU consumption of critical raw materials over South African efforts to meet their development objectives through establishing a PGM/fuel cell hub. Incorporating this political risk discourse into criticality assessments ultimately reflects the colonial legacies which infuse contemporary extractive industry practices—even those nominally oriented toward resourcing decarbonized futures.

Conclusion: Provincializing criticality assessments

Critical raw materials assessments conducted by the wealthy Northern economies respond to the growing componential complexity and mineral intensity of emerging technologies, but ultimately to anxieties over the security of raw material supply. While there is no harmonized definition of criticality in the existing literature (Habib & Wenzel, 2016; Mancini et al., 2018), what all extant criticality assessments share is (1) a concern with resource use on the part of wealth, resource-importing Northern economies, and (2) the use of "political risk" metrics to determine supply risk and thus criticality. In this chapter, I have examined dominant approaches to critical raw materials assessment from the perspective of Science and Technology Studies, and the literature on "resource-making" and "valuation." I have argued that incorporating measures of political risk into criticality assessments reflects a Eurocentricity and reproduces problematic colonial discourses which portray expressions of postcolonial sovereignty as aberrant or deviant behavior. The Fraser Institute's Policy Potential Index (PPI) and the World Bank's World Governance Indicators (WGIs) have been the dominant vehicles for the inclusion of political risk discourse into criticality assessments.

This chapter has also examined cobalt/lithium in DRC and PGMs in South Africa, in order to argue that current models of criticality assessment need to be "provincialized" and decoupled from their commitment to the concerns of the Global North while highlighting that current extraction of "critical" materials crucial to decarbonized energy futures is embedded in exploitative postcolonial relationships. The cases from DRC focused on stability clauses, a key artifact of the colonially derived system of international investment law, work to transform sovereign postcolonial states into private contracting parties subordinated to the desires extractive industry corporations. By expressing concerns about supply risk of critical raw materials in the language of "political risk," criticality assessments implicitly endorse a residual colonial system of international law which privileges stability over postcolonial sovereignty.

The DRC cases also highlighted the manner in which attempts to secure royalties on the part of postcolonial nations is presented as a threat to supply risk, given that greater royalty rates can undermine net present value assessments, and destabilize attempts at resource-making. Yet, if any serious attempt to meet the 1.5°C Paris targets took place, demand for cobalt and lithium should be sufficient to allow investors to make critical raw material deposits valuable, even in the face of high royalty rates. It is, therefore, apparent that there is a need to provincialize criticality assessments and ensure that critical raw materials assessments serve resource imaginaries in which postcolonial nations like DRC do not merely serve as sites of extraction for the wealthy North, but participate fully in the transition to decarbonized energy and transport infrastructure. These arguments were further developed in relation to South Africa and PGMs, where attempts to secure decent conditions for mineworker are reduced to political risk that poses a threat to supply (and to resource-making and valuation), and these measures of political risk are once again incorporated into EU and other criticality assessments. As Mabanza (2018) highlights, wealthy

Northern economies are at pains to ensure that international investment law and economic partnership agreements *de-risk* supply of critical raw materials, but these global political-economic inequalities are naturalized, rather than appreciated as expressions of the extractive industries' coloniality. Provincialized critical raw materials assessments, then, would also need to take account of the legal and treaty networks which work to de-risk supply for some, but not other, players in the global extractive economy.

References

Anghie, A. (2007). *Imperialism, sovereignty and the making of international law*. Cambridge: Cambridge University Press.

Australian Mining. (2019). *AVZ confirms Manono as world's largest lithium resource*. May 8 2019. https://www.australianmining.com.au/news/avz-confirms-manono-as-worlds-largest-lithium-resource/ (Accessed August 9, 2019).

Bessarabov, D., van Niekerk, F., van der Merwe, F., Vosloo, M., North, B., & Mathe, M. (2012). Hydrogen infrastructure within HySA national program in South Africa; road map and specific needs. *Energy Procedia, 29*, 42–52.

Bond, P. (2013). Debt, uneven development and capitalist crisis in South Africa: From Moody's macroeconomic monitoring to Marikana microfinance mashonisas. *Third World Quarterly, 34*(4), 569–592.

Bush, S. S. (2017). The politics of rating freedom: Ideological affinity, private authority, and the Freedom in the World ratings. *Perspectives on Politics, 15*(3), 711–731.

Chakrabarty, D. (2000). *Provincializing Europe: Postcolonial thought and historical difference*. Princeton, NJ: Princeton University Press.

Cimprich, A., Young, S. B., Helbig, C., Gemechu, E. D., Thorenz, A., Tuma, A., et al. (2017). Extension of geopolitical supply risk methodology. Characterization model applied to conventional and electric vehicles. *Journal of Cleaner Production, 162*, 754–763.

Clowes, W. (2018). *Congo triples levy on cobalt with strategic minerals decree*. 3 December 2018. https://www.bloomberg.com/news/articles/2018-12-03/congo-triples-levy-on-cobalt-with-strategic-minerals-decree (Accessed August 9, 2019).

Cobham, A., Janský, P., & Meinzer, M. (2015). The Financial Secrecy Index: Shedding new light on the geography of secrecy. *Economic Geography, 91*(3), 281–303.

De Ville, F., & Siles-Brugge, G. (2015). The transatlantic trade and investment partnership and the role of computable general equilibrium modelling: An exercise in 'managing fictional expectations'. *New Political Economy, 20*(5), 653–678.

Dumit, J. (2014). Writing the implosion: Teaching the world one thing at a time. *Cultural Anthropology, 29*(2), 344–362.

Emel, J., & Huber, M. T. (2008). A risky business: Mining, rent and the neoliberalization of 'risk'. *Geoforum, 39*, 1393–1407.

European Commission. (2017a). *Study on the review of the list of critical raw materials: Critical raw materials factsheets*. Brussels: Directorate-General for Internal Market, Industry, Entrepreneurship and SMEs.

European Commission. (2017b). *Study on the review of the list of critical raw materials: Non-critical raw materials factsheets*. Brussels: Directorate-General for Internal Market, Industry, Entrepreneurship and SMEs.

Ferguson, N., & Schularick, M. (2006). The empire effect: The determinants of country risk in the first age of globalization, 1880-1913. *The Journal of Economic History*, 66(2), 283–312.

Ferry, E. E., & Limbert, M. (2008). Timely assets. In E. E. Ferry & M. Limbert (Eds.), *Timely assets: The politics of resources and their temporalities* (pp. 3–24). Santa Fe: School for Advanced Research Press.

Forslund, D. (2015). Briefing on the report 'The Bermuda Connection: Profit shifting, inequality and unaffordability at Lonmin 1999-2012'. *Review of African Political Economy*, 42, 657–665.

Gemechu, E., Helbig, C., Sonnemann, G., Thorenz, A., & Thuma, A. (2017). Import-based Indicator for the geopolitical supply risk of raw materials in life cycle sustainability assessments. *Journal of Industrial Ecology*, 20(1), 154–165.

Gilbert, P. R. (2018). Sovereignty and tragedy in contemporary critiques of investor state dispute settlement. *London Review of International Law*, 6(2), 211–231.

Gilbert, P. R. (2019). Bangladesh as the 'next frontier'? Positioning the nation in a global financial hierarchy. *Public Anthropologist*, 1(1), 62–80.

Gilbert, P. R. (2020a). Expropriating the future: Turning ore deposits and 'legitimate expectations' into assets. In K. Birch & F. Muniesa (Eds.), *Turning things into assets: New lineaments in the study of technoscientific capital*. Cambridge MA: MIT Press.

Gilbert, P. R. (2020b). Speculating on Sovereignty: 'Money mining' and corporate foreign policy at the extractive industry frontier. *Economy and Society*. https://doi.org/10.1080/03085147.2019.1690255.

Giurco, D., Dominish, E., Florin, N., Watari, T., & McLellan, B. (2019). Requirements for minerals and metals for 100% renewables scenarios. In S. Teske (Ed.), *Achieving the Paris climate agreement: Global and regional renewable energy scenarios with non-energy GHG pathways for +1.5C and +2C* (pp. 437–457). Amsterdam: Springer.

Gutstein, D. (2014). *Harperism: How Stephen Harper and his think tank colleagues have transformed Canada*. Toronto: James Lorimer.

Habib, K., Hamelin, L., & Wenzel, H. (2016). A dynamic perspective of the geopolitical supply risk of metals. *Journal of Cleaner Production*, 133, 850–858.

Habib, K., & Wenzel, H. (2016). Reviewing resource criticality assessment from a dynamic and technology specific perspective—Using the case of direct-drive wind turbines. *Journal of Cleaner Production*, 112, 3852–3863.

Hobson, P. (2019). *Platinum week: Oversupply looms over investor bets on platinum rebound*. 13 May 2019. https://uk.reuters.com/article/platinum-week-price-idUKL5N22P3U1 (Accessed August 9, 2019).

Hollins, O., & Fraunhofer, I. S. I. (2013). *Study on critical raw materials at EU level: Final report for DG Enterprise and Industry*. Aylesbury: Oakdene Hollins.

Hossain, K. (1983). Introduction. In K. Hossain & S. R. Chowdhury (Eds.), *Permanent sovereignty over natural resources in international law* (pp. ix–xx). London: Frances Pinter.

Hu, T. (2019). *China's fuel cell push not seen stimulating short-term platinum demand*. 10 June 2019. https://www.spglobal.com/marketintelligence/en/news-insights/latest-news-headlines/52231420 (Accessed August 9, 2019).

Huber, M. (2018). Resource geography II: What makes resources political? *Progress in Human Geography*, 43(3), 553–564.

Hume, N. (2019). *'No more debate' on DRC mining code*. Financial Times. 5 February 2019 https://www.ft.com/content/713c0686-2952-11e9-88a4-c32129756dd8 (Accessed August 9, 2019).

Ioannidou, D., Heeren, N., Sonnemann, G., & Habert, G. (2019). The future in and of criticality assessments. *Journal of Industrial Ecology*, 23(4), 751–766.

Jamasmie, C. (2018). *DRC miners form new body to fight mining code.* 23 August 2018. https://www.mining.com/drc-miners-form-new-body-fight-mining-code/ (Accessed August 9, 2019).

Kama, K. (2019). Resource-making controversies: Knowledge, anticipatory politics and economization of unconventional fossil fuels. *Progress in Human Geography,* https://doi.org/10.1177/0309132519829223.

Kantor, M. (2015). Indirect expropriation and political risk insurance for energy projects. *Journal of World Energy Law and Business, 8*(2), 173–198.

Kobrin, S. J. (1982). *Managing political risk assessment: Strategic response to environmental change.* Berkeley, CA: University of California Press.

Kornberger, M., Justesen, L., Madsen, A. K., & Mouritsen, J. (2015). Introduction: Making things valuable. In M. Kornberger, L. Justesen, A. K. Madsen, & J. Mouritsen (Eds.), *Making things valuable* (pp. 1–17). Oxford: Oxford University Press.

Lawhon, M., Ernstson, H., & Silver, J. (2013). Provincializing urban political ecology: Towards a situated UPE through African urbanism. *Antipode, 46*(2), 497–516.

Linarelli, J., Salomon, M., & Sornarajah, M. (2018). *The misery of international law: Confrontations with injustice in the global economy.* Oxford: Oxford University Press.

Mabanza, B. (2018). Imperialist raw materials strategies in EU politics. In M. Grimm, J. Krameritsch, & B. Becker (Eds.), *Business as usual after Marikana: Corporate power and human rights* (pp. 334–351). Johannesburg: Jacana.

Mahamba, F., & Cocks, T. (2018). *Congo mining code regulations signed into law—Aides.* Reuters. Mining.Com. 10 June 2018. https://www.mining.com/web/congo-mining-code-regulations-signed-law-aides/ (Accessed August 9, 2019).

Mancini, L., Benini, L., & Sala, S. (2018). Characterization of raw materials based on supply risk indicators for Europe. *International Journal of Life Cycle Assessment, 23,* 726–738.

Metayer, M., Breyer, C., & Fell, H.-J. (2015). *The projections for the future and quality in the past of the World Energy Outlook for solar PV and other renewable energy technologies.* Berlin: Energy Watch Group.

Mining Journal. (2018). *DRC declares Coltan 'strategic'.* 4 December 2018. https://www.mining-journal.com/politics/news/1352362/drc-declares-cobalt-%E2%80%98strategic%E2%80%99 (Accessed August 9, 2019).

Mossallam, M. (2015). *Process matters: South Africa's experience exiting its BITs.* Global Economic Governance Programme WP 2015/97, Oxford.

Muniesa, F. (2011). A flank movement in the understanding of valuation. *The Sociological Review, 59,* 24–38.

Muniesa, F., Doganova, L., Ortiz, H., Pina-Stranger, Á., Paterson, F., Bourgoin, A., et al. (2017). *Capitalization: A cultural guide.* Paris: Mines Paris Tech.

Nansai, K., Nakajima, K., Kagawa, S., Kondo, Y., Shigetomi, Y., & Suh, S. (2015). Global mining risk footprint of critical metals necessary for low-carbon technologies: The case of neodymium, cobalt and platinum in Japan. *Environmental Science & Technology, 40,* 2022–2031.

National Research Council. (2008). *Minerals, critical minerals, and the U.S. Economy.* Washington, D.C.: National Academies Press.

Norton Rose Fulbright. (2018). *Major changes to the mining code of the Democratic Republic of Congo.* https://www.nortonrosefulbright.com/en/knowledge/publications/07ca4707/major-changes-to-the-mining-code-of-the-democratic-republic-of-congo (Accessed August 9, 2019).

Oil Change International and IEEFA. (2018). *Off track: How the international energy agency guides energy decisions towards fossil fuel dependence and climate change.* Washington, DC: Oil Change International.

Onstad, E. (2019). *Exclusive: Bosch goes for platinum-light fuel cells*. 13 May 2019. https://www.reuters.com/article/us-platinum-week-bosch-fuelcells-exclusi-idUSKCN1SJ0FG (Accessed August 9, 2019).

Overland, I. (2019). The geopolitics of renewable energy: Debunking four emerging myths. *Energy Research & Social Science, 49*, 36–40.

Pollet, B. G., Pasupathi, S., Swart, G., Mouton, K., Lotoskyy, M., Williams, M., et al. (2014). Hydrogen South Africa (HySA) systems competence centre: Mission, objectives, technological achievements and breakthroughs. *International Journal of Hydrogen Energy, 39*(2014), 3577–3596.

Rajak, D. (2014). Corporate memory: Historical revisionism, legitimation and the invention of tradition in a multinational mining company. *Political and Legal Anthropology Review, 37*, 259–280.

Rangold Resources. (2018). *Randgold resources Ld announces new mining industry body in DRC* 23 August 2018 https://www.accesswire.com/510637/Randgold-Resources-Ld-Announces-New-Mining-Industry-Body-in-DRC (Accessed September 8, 2018).

Richardson, T., & Weszkalnys, R. (2014). Introduction: Resource materialities. *Anthropological Quarterly, 87*, 5–30.

Ritchken, E. (2018). The gold of the 21st century: A vision for South Africa's platinum group metals. In S. Valiani (Ed.), *The future of mining in South Africa: Sunset or sunrise?* (pp. 66–120). Mapungubwe Institute for Strategic Reflection: Johannesburg.

Sainsbury, P. (2012). *What impact will Marikana unrest have on platinum price?* 19 August 2012 https://materials-risk.com/impact-marikana-unrest-platinum-price/ (Accessed August 9, 2019).

Sanderson, H. (2018). *Lithium prices to fall by 45%, Morgan Stanley says*. Financial Times 26 February 2018. https://www.ft.com/content/66012fe2-1ae1-11e8-aaca-4574d7dabfb6.

Sanderson, H. (2019). *Plummeting cobalt price takes toll on Democratic Republic of Congo*. Financial Times. 24 June 2019. https://www.ft.com/content/3317dc52-933a-11e9-aea1-2b1d33ac3271 (Accessed August 9, 2019).

Schlemmer, E. (2015). Overview of South Africa's bilateral investment treaties and investment policy. *ICSID Review, 31*(1), 167–193.

Styve, M. D. (2019). *From Marikana to London: The anti-blackness of mining finance*. PhD ThesisUniversity of Bergen.

Tsing, A. (2000). Inside the economy of appearances. *Public Culture, 12*(1), 115–144.

Watari, T., McLellan, B., Ogata, S., & Tezuka, T. (2018). Analysis of potential for critical metal resource constraints in the international energy agency's long-term low-carbon energy scenarios. *Minerals, 8*(156), 1–34.

Whitehouse, D. (2019). *South Africa miners need price swings just to stay afloat*. 26 February 2019. https://www.theafricareport.com/383/south-africa-miners-need-price-swings-just-to-stay-afloat/ (Accessed August 9, 2019).

Environmental impacts of mineral sourcing and their impacts on criticality

Benjamin C. McLellan

Graduate School of Energy Science, Kyoto University, Kyoto, Japan

Introduction

Mineral demand has been continuously and rapidly increasing (McLellan, Corder, Giurco, & Ishihara, 2012; McLellan, Diniz Da Costa, & Dicks, 2012). The requirement of minerals for supporting clean energy transitions has brought significant additional demand for many minerals that were previously of only minor importance, as well as expanding the need for some of the bulk minerals for fundamental energy infrastructure such as transmission and distribution grids (Watari, McLellan, Ogata, & Tezuka, 2018) (see Wang et al. and Zepf within this book). These pressures from additional consumption and anticipated further growth if clean energy transitions are to be achieved in reality, as well as factors such as political intervention in supply, have led to concerns over the future availability of certain minerals and related studies arguing for or against the likelihood of supply constraints—for example, rare earths (McLellan, Corder, & Ali, 2013), lithium (Grosjean, Miranda, Perrin, & Poggi, 2012), and platinum (Alonso, Field, & Kirchain, 2012; Harvey, 2018). In conjunction with these recent concerns, and in part as a response, there have been an increasing number of studies of "mineral criticality" including the development and refinement of methodologies as well as a variety of studies on selected minerals, technologies, and countries (Erdmann & Graedel, 2011; Graedel & Reck, 2016) (see Gilbert, Wang et al., Zepf within this book). Notable in the progress of many of these methodologies has been a gradual expansion of the original scope of evaluation from indicators of economic importance, geological and political stability as factors of risk to reliable and affordable supply of important minerals, to include environmental risk-related factors (Watari et al., 2019) (see Penaherrera and Pehlken within this book). This evolution follows similar developments in the parallel field of energy security assessment, through which governments seek to evaluate the risks to continuity of supply of safe, accessible, and affordable energy. These environmental risks in the supply chain, in addition to the social and governance risks, can be seen as potentially important to future availability. This chapter will expand on this

The Material Basis of Energy Transitions. https://doi.org/10.1016/B978-0-12-819534-5.00007-6

particular area of minerals criticality, with a specific focus on the localization and specification of environmental impacts and risks, as opposed to the globalization and generalization that is still typical.

Environmental impacts—Scope and data

When environmental impacts and risks are being considered within the framework of critical minerals, it is important to consider the scope of the assessment, as this gives crucial insights into the level of applicability of results, and the gaps that may need to be evaluated further. Broadly speaking, environmental impacts can be considered from two perspectives—the global or lifecycle perspective or the local perspective. While these are not mutually exclusive approaches, they tend to apply different methodologies, require different data and are likely to be applicable to different levels of assessment. Both of these perspectives are introduced below, with a comparison following.

The lifecycle perspective

The consideration of minerals as material inputs can already be considered as an application of the methodology of lifecycle assessment (LCA)—an ISO standardized methodology that fundamentally takes a cradle-to-grave approach to evaluating the environmental impacts of a process, service, or product. Within the LCA approach, the inputs and outputs of material and energy, at each stage of the supply chain of the product, are inventoried and then the subsequent environmental performance is evaluated by the characterization and estimation of likely impacts associated with each of the particular inflows and outflows. For example, the LCA of photovoltaic (PV) power generation would consider the direct material requirement for a PV panel, then work back through the manufacturing of the panel, the manufacturing of the components, back to the production of the primary minerals extraction—considering the inputs and outputs (emissions as well as products) at each stage. Material requirements would be aggregated into natural resource depletion indicators, emissions would be variously characterized into specific environmental impact categories on the basis of their observed impact on the environment, and eventually those impact categories may be weighted and accumulated into one or more compound indicators (Huijbregts et al., 2016). The LCA approach is very useful, particularly for comparing globally relevant impacts and globally averaged supply chains. It is, however, somewhat more limited when specific supply chains and localized impacts are being considered—particularly for countries outside of Europe, the United States, and Japan, which are the main countries and regions that have available data and locally derived characterization models/methods. Most of the mineral criticality methods that have utilized environmental impacts as part of the assessment so far have incorporated LCA-type indicators, often following the precedents of leading groups within the field of criticality assessment (Graedel et al., 2012). Others have used, for example, the total material requirement (TMR) (Watari et al., 2019), which is an offshoot indicator from LCA that combines all direct and indirect material inputs to a process into a single mass value. It can be argued that the use of generalized LCA indicators such as ReCiPe points or TMR may be useful for evaluating criticality at the global scale (and even then the error due to lack of geographically explicit data may be high), but may not be as clearly justifiable when considering national-level criticality.

Data availability, accuracy, and age are significant factors that affect the uncertainty of all LCA studies, but are likely to be exacerbated in the case of minerals due to a number of inherent characteristics of minerals supply chains. First, mineral deposits vary significantly in the grade (concentration of useful mineral contained in the ore) and associated desirable and undesirable elements. This leads to a variety of mineral processing routes, which can have large differences in their applicability as well as their environmental impact (Norgate & Jahanshahi, 2010). Moreover, the ore characteristics can change overtime as economic and technological factors can make marginal ore viable for extraction. Second, many of the current mineral deposits and well-defined reserves are in the developed countries, while the deposits yet to be exploited are in countries that generally have lower domestic capacity for exploration and regulation of reported data, which may affect the estimations of future calculations on the specific routes to be applied (including something as relatively simple as the depth and transportation requirement for accessing these deposits). Third, as described above, much of the LCA data has been developed on the basis of North American and European conditions, whereas the characterization factors may not be representative in emerging supply countries. While all of these factors are not insurmountable, and while the LCA approach probably represents the most readily applicable method for global criticality studies especially, it is important to note that there are alternatives that might be more applicable in other situations.

Local environmental impact perspective

While LCA considers the full supply chain and combines the impacts across the life cycle, the reliance on generalized characterization methods and limited data can disguise the direct impacts on the environment at each stage and reduce the possibility of identifying thresholds or limits to reversible damage. This is in contrast to the process of accumulating LCA data itself, which often relies on local measurements and in the case of minerals can rely on the data of specific ore processing operations. The local environmental impacts of minerals and their supply chains—from the mine through to the final manufacturing of the product (and potentially through to recycling or disposal)—are generally assessed operation-by-operation and independently of the other elements of the supply chain. Approaches for the evaluation of localized impacts are included in frameworks such as Environmental Impact Assessment (EIA), which are often state regulated and can vary widely with regards to their requirements and application. EIA is a process by which the anticipated environmental impacts of a proposed project (e.g., a mining operation or power plant) are estimated and reported on the basis of engineering designs prior to the regulatory acceptance of the project. These are typically reports made by specialist consultants, covering a broad range of environmental impacts, but typically with a local focus and specific to the project and sites being investigated. They are often made available for public comment, and should include consideration alternatives to mitigate environmental impacts. EIA as a common regulatory process is a useful data source, particularly given that the reports for such processes are often made available for public comment. EIA typically provides a baseline evaluation and an estimate of potential impacts for an operation on its local environment. EIA usually involves modeling of atmospheric, noise and hydrological pollutant transmission and the likelihood and magnitude of impact on surrounding environment and population. Good EIAs typically include a consideration of the cumulative impacts of additional operations in environments that have other proposed or current industrial operations. This is particularly relevant in the case of mines, which can often be found in close proximity to each other due to regional geology. It is becoming common to also have Social Impact Assessments on new operations, but during some periods of methodological and regulatory development, SIAs have been included as subsections of EIAs, thus also providing useful socioeconomic data.

EIAs often contain or are accepted as a basis for the production of an Environmental Management Plan (EMP) that will provide operational emissions limits and regulatory monitoring and reporting requirements. This regulatory reporting as well as, more recently, voluntary sustainability reporting is another important data resource for examining the local impacts of critical minerals provision. In fact, there are a host of existing reporting requirements in many countries that could provide useful data for understanding the local impacts and progress across the lifetime of a project (McLellan, 2014).

Within the context of mineral provision for clean energy technologies, the local environmental impacts are likely to be important factors for a number of reasons. First, the direct impact on the environment is related to the local environment, the mineralogy, processing methods, and regulated pollution controls. This means that there is potential for differences in the relative environmental impact for producing the same mineral at different locations. Mines are fixed in their location due to the geological presence of ore, so the specific environmental impacts are also fixed. One clear example of differences in local mining impacts is the production of lithium from brines in Chile (concentrated lithium-containing groundwater is pumped from the ground and evaporated to extract the lithium-containing minerals, before further processing) as opposed to the production from ores in Australia (a more typical mining operation that extracts hard-rock ore for crushing and further processing)—there is an inherent difference in potential environmental impact due to the different processes leading to different emissions. But there can also be different impacts due to the host environment itself—for example, water consumption in a high rainfall area versus a low rainfall area will result in significantly different pressures on the environment (McLellan, Corder, et al., 2012; McLellan, Diniz Da Costa, et al., 2012). There are indications that much of the potential for new mines to provide for the expected increase in demand for critical minerals have geological properties and are located in environmentally (and socially) challenging locations (Valenta, Kemp, Owen, Corder, & Lèbre, 2019). The potential for environmental impacts can lead to loss of social license to operate, resulting in regulatory restriction or blocking of production, which can exacerbate potential supply risks. In most cases, it is local environmental emissions and impacts that are likely to cause shutdowns, rather than global emissions.

There are certain environmental impacts that are clearly not localized—for example, emissions that contribute to climate change. There are also emissions that are both locally and globally relevant, such as ozone and ozone-depleting substances that can have different impacts as they are transported from the surface of the Earth to higher in the atmosphere. LCA takes this into account, but ultimately comes up with a set of minimal aggregated indicators for comparison. In the case of local environmental impacts, there remains an important issue of how to compare alternative sites or alternative supply chains with multiple environmental issues (e.g., is a site that uses more water but emits less SO_x better than a site that does the opposite?). Moreover, in the context of criticality, there is the additional need to clarify the likelihood of a certain mine being allowed to operate as opposed to others—with many of the main drivers for operation being economic rather than environmental.

Energy and environmental impacts

When considered in the context of renewable or clean energy and the reduction of global emissions, local environmental emissions also need to be weighed against the ability to reduce CO_2 emissions by utilizing a preferred supply chain. In many cases, it is the energy supply to a mine or mineral process that will determine the lifecycle reduction of greenhouse gas emissions, but this will be affected specifically by the energy mix used by the operation. The energy mix for the mine can be dependent on broader state or national energy mixes if the mine is grid connected, or the operation may supplement or even be independent of the grid, therefore, giving the operator control over the energy mix selection. In support of this latter approach, a large proportion of mines are located in remote areas and many have onsite power generation to ensure constant supply, particularly in areas where grid electricity is inconsistent. Such remoteness, and the typical availability of land in the surrounds of a mine that this often implies, can be of benefit if mine owners choose to install renewable energy (such as PV) (McLellan, Corder, et al., 2012; McLellan, Diniz Da Costa, et al., 2012). This approach has been becoming popular both for mines during their operation phase as well as part of postmine rehabilitation. Further down the supply chain, the ability to install renewable energy onsite is likely to reduce due to proximity to workforce and infrastructure in towns, but there is potential to shift operations to other countries that have lower emissions energy mixes, as has been somewhat of a trend with aluminum smelters (McLellan, 2011).

It is not just the energy mix, but also the energy efficiency or energy intensity of mineral supply chains that needs to be considered. The amount of energy consumed, and subsequent global or local emissions, is related to the same geological and mineralogical factors, as well as the technology utilized for extraction and processing—with some clear interconnections between the geology, mineralogy, and the technology for mining and processing options that is chosen (Mudd, 2007a, 2007b, 2010). It should be noted that, in most of the energy scenarios that consider mineral requirements as a constraint or output, there is a broad lack of consideration of expanded energy consumption for the minerals sector as mineral requirements increase and also a neglect of potential effects of declining grade and deeper mining that can exacerbate this increase. There are only few exceptions to this, and those studies have only addressed generalized minerals sectors, largely at the global scale (Tokimatsu et al., 2018) with some advances for specific countries (Elshkaki, 2019).

Changing environmental impacts and mitigation

Whether considered from a lifecycle perspective or a local environmental perspective, the important factor to consider is that the clean energy transition will lead to shifts in environmental impacts. LCA tries to ensure that such shifts that might occur from one stage of the life cycle to another are still incorporated in the overall environmental burden (see Penaherrera and Pehlken within this book). As previously mentioned, differential environmental impacts can be used to determine the ideal supply chain of minerals as well. This section briefly outlines some of the potential shifts in environmental impacts at each stage and along the supply chain.

Mines

Mines are geographically fixed due to the geological occurrence of economically viable mineral deposits. This makes the local factors around current and future mines a particularly important consideration. Typically, mining operations suffer from a number of key environmental concerns, chiefly:

- Water usage and pollution
- Dust and other atmospheric emissions
- Landform and vegetation change

Other notable impacts such as on flora and fauna, noise, and light pollution are, of course, important to the local area and community, but will not be focused on here.

Water is a very important issue in mines across the world. Even without considering the first stages of mineral processing that would typically produce a mineral concentrate for metalliferous ores, there is a strong dependence on water consumption at mines, for example, in the suppression of dust for safety and environmental purposes. Mines, particularly as mining goes deeper underground in order to access more ore, often reach below the water table. Some mineral deposits themselves—particularly in the case of coal—often form part of the local aquifer system, with mining having a potential significant effect on the natural flow of groundwater. Moreover, particular minerals—the most relevant example being lithium—are extracted from brine deposits, making the extractive process a direct water impact. Water management on mines has long been an issue of key importance to mining companies—stopping mine flooding being a key operational concern, while reducing water loss is an operational cost mitigation focus (Ihle & Kracht, 2018). Water content in tailings dams can also be a key issue for mine safety, as heavy rainfall can be one contributor in dam failure that can ultimately lead to catastrophic social and environmental consequences (Burritt & Christ, 2018). The impact of mines on potential water availability for the surrounding community or other water-dependent industries such as agriculture can be a key flash-point issue for the loss of social license to operate. This impact may be due to the degradation of water quality, quantity, or stability of supply to the relevant stakeholders. Even the potential threat of loss of water access may be enough to cause community backlash and failure of new mines to gain regulatory approval. This type of water issue is prominent in areas with low rainfall and/or high-water extraction rates for preexisting or socially prioritized industries.

Water contamination or pollution has been a long-standing issue with mining operations. In the operational phase, releases of chemical contaminants (although more frequently associated with the mineral processing stages of the life cycle) or even of relatively inert sediment load into rivers, lakes, or oceans has resulted in the death of aquatic species and sometimes terrestrial species that prey on them (including humans). There have been numerous historical examples of legacy mines that have resulted in acid mine drainage (AMD) with ongoing treatment required for decades if not perpetually (Tuazon & Corder, 2008). In such cases, high rainfall areas can be adversely affected as the risk of uncontained mine runoff increases.

Landform change is an inevitable part of most types of mining—ore is typically extracted by directly removing it from the ground, thereby leaving large voids in or under the ground and often accumulating the waste material as hills or mounds beside these holes. As mining has been shifting from underground to open cut and also toward lower grade and deeper ores, there has been a notable shift in the total mine waste production (Mudd, 2007a, 2007b). Legacy mines can have a variety of impacts, many of which are highly affected by the changes in landform and in vegetation cover (Worrall, Neil, Brereton, & Mulligan, 2009). In some cases, there have been constructive moves to utilize the new landforms—for example, for renewable energy (PV panels and wind turbines on postmined land or even floating PV on water-filled voids), or for pumped hydropower storage (McLellan, Choi, Ghoreishi-Madiseh, & Hassani, 2018).

Dust and other atmospheric emissions are highly important across the supply chain, and there is a good argument for being able to sum these emissions across the life cycle, particularly emissions that have effects over a longer distance and timescale (up to global). However, the concentration of these emissions is the key factor in determining health and environmental impacts, and concentration is not an additive factor across disparate geographic locations. It is notable that atmospheric emissions that are visible, malodorous, or cause a physical reaction (e.g., stinging eyes) are most likely to cause societal outrage, but they may not necessarily be the most damaging to health or the environment. Dust emissions are almost guaranteed to occur in most forms of mining—and as mentioned earlier, dust suppression is one of the key areas of water use in many mines. In order to extract minerals, mines will use explosives or tools to break ore away from the ore body, and then transport the partially fragmented rock by truck or conveyor to where it will be further crushed in order to liberate the useful mineral from the waste rock. The breaking of rock is essential in most cases, and transporting it anywhere will typically lead to dust emissions from the ore as well as from the roads it is transported on. Heavy equipment also tends to be run on diesel fuel, which adds to atmospheric emissions.

Water usage and pollution, and landform or land use change are highly locally relevant issues that are not particularly well dealt with by LCA. A cumulative value of water consumption (for example, a water footprint) is typically not able to tell the level of water stress in the region where mining is occurring or the cumulative or relative effects that water use and contamination might have in a locality—and combining these into a lifecycle indicator is not easy. Likewise with landform change and land use change, the specific effects are highly localized, meaning that generalized quantitative factors such as the cumulative land disturbance or weighted sums of land use change are overly simplistic. As with biodiversity loss, LCA tools have made some progress, but it is still difficult to justify a summation across the life cycle when there are potentially quite different geographical areas involved. Dust and atmospheric emissions, while having a better argument for summation of total emissions, are still most relevant at the local scale, as concentration in the air and the transmission of the pollutant to local receptors (such as the community) are the key determinants of impact. LCA impact assessment tools have generalized models for these, but they suffer from accuracy concerns outside the areas where they were developed and tested.

The critical factor to consider with these impacts is that they will also change according to the type of mineral that is being mined. For example, coal mining will produce coal dust, but copper mining will produce dust that inherently contains copper. Though both are dust, there will be a differential effect on the environment and on human health in the regions impacted. This is a crucial factor to be considered with regards to the potential supply risks due to environmental impacts in the criticality assessment of minerals for clean energy. As the energy mix moves from coal toward renewables, there will be a shift in the location of mining (due to geology) and the type of emissions (due to the type of mineral being mined). So, while emissions in coal mining regions may be mitigated, copper mining regions and lithium mining regions might experience degradation of their environment through increased emissions. Weighing the ultimate impact of such a shift in geographical location and type of emissions remains a challenge.

Processing, extraction, smelting, and refining

The downstream phase of taking the mined mineral ore and producing a metal or useful commodity from it are the phases that tend to be most energy and chemical consuming, whilst also creating more added economic value. As ores become more complex, lower grade, and contain unwanted by-products such as arsenic, their treatment also becomes more environmentally impactful. Many of the materials that are used in renewable energy technologies occur as by-products or coproducts with other metals. The drive to extract the minor elements can be beneficial from the perspective of reducing waste (potentially, by removing more elements from the tailings) but can also require more processing. These ongoing changes in extractive emissions need to be monitored to update criticality assessments effectively.

Demand side

While it is inherent in the theme of this book, it is worth noting that there will be a trade-off of the emissions at the demand side with the remainder of the supply chain. While renewable energy technologies can replace fossil fuels in power generation—sometimes at the same location (Chapman, McLellan, & Tezuka, 2018) —there will be direct environmental benefits, as well as the supply chain effects. Of course, the impact on greenhouse gas emissions is the target of clean energy transitions, but the reduction of other emissions—such as particulates, heavy metals, and gaseous emissions—can be of equal or larger importance, and often have a more direct and immediate impact on the lives of host communities. Other clean energy applications, such as electric or hydrogen vehicles can also show significant co-benefits, particularly in mitigating emissions in cities. The balance of detrimental emissions increases at the sites of mining and processing, refining, or smelting need to be weighed up against the benefits of reducing emissions elsewhere.

Waste and recycling

Some of the impacts that need to be further considered in the life cycle of renewable energies specifically are the reduction of wastes and the potential need to recycle end-of-life equipment such as lithium-ion batteries and PV panels (see Goddin, Chapter 13, within this book). As these technologies have been rapidly increasing in demand and subsequently in production, there is a growing stock of equipment that will start to come out of operation in the next decade. This poses an environmental challenge in that the equipment needs to be effectively treated, recycled, or reused where possible and finally disposed of where necessary. For lithium-ion batteries, one of the critical issues has been that recycling is low overall, but what recycling does occur has been for the electrode materials (e.g., cobalt) rather than for the lithium itself (see Weil et al. within this book). While this is commercially driven at present, it could be expected that recycling will eventually be driven by resource security concerns, as well as the potential to reduce the environmental impact of the supply chain by reducing reliance on primary mined content (Watari et al., 2018). For PV panels, the potential for a surge in retired panels has been causing concern in some countries, with no consistent and efficient existing recycling systems in place. Moreover, the different types of panels—thin film vs silica cells—make the materials contained and their recovery techniques more complex. Currently for many metals, recycling from final renewable energy products does not give any great energy efficiency benefit when compared with primary metals, so it is other drivers that are currently more prominent in considering recycling.

The increase in mining for these energy minerals will also likely lead to an increase in tailings. The need for safe long-term storage of tailings has been a concern for a long time—particularly in the recent years due to notable dam failures. There is potential for reenvisioning these tailings as either a useful material input for value-added products (such as cement or geopolymer) or maintaining them for the situation in which extraction of minor elements may become economically feasible with advances in technology. In either case, dealing with this mine waste is an important concern, and there has been a conscious shift in some sectors to aim for drastic reductions in waste through such by-products.

Conclusions

Environmental impacts from the extraction of minerals for the clean energy transition, and the integration of these into evaluation methods, are an area of increasing research efforts in the field of critical minerals. There are many potential environmental impacts, but there are two broad perspectives on them—the local and the global or generalized perspectives. Techniques such as LCA are currently used in evaluating mineral criticality as an indicator of the potential environmental risk associated with certain metal supply chains. Local impacts are often more relevant—particularly for mining, which is geographically fixed due to geology. Therefore, further methods need to be produced to better incorporate these localized impacts.

References

Alonso, E., Field, F. R., & Kirchain, R. E. (2012). Platinum availability for future automotive technologies. *Environmental Science & Technology*, *46*, 12986–12993. https://doi.org/10.1021/es301110e.

Burritt, R. L., & Christ, K. L. (2018). Water risk in mining: Analysis of the Samarco dam failure. *Journal of Cleaner Production*, *178*, 196–205. https://doi.org/10.1016/j.jclepro.2018.01.042.

Chapman, A. J., McLellan, B. C., & Tezuka, T. (2018). Prioritizing mitigation efforts considering co-benefits, equity and energy justice: Fossil fuel to renewable energy transition pathways. *Applied Energy*, *219*, 187–198. https://doi.org/10.1016/j.apenergy.2018.03.054.

Elshkaki, A. (2019). Material-energy-water-carbon nexus in China's electricity generation system up to 2050. *Energy*, 116355. https://doi.org/10.1016/j.energy.2019.116355.

Erdmann, L., & Graedel, T. E. (2011). Criticality of non-fuel minerals: A review of major approaches and analyses. *Environmental Science & Technology*, *45*, 7620–7630. https://doi.org/10.1021/es200563g.

Graedel, T. E., Barr, R., Chandler, C., Chase, T., Choi, J., Christoffersen, L., et al. (2012). Methodology of metal criticality determination. *Environmental Science & Technology*, *46*, 1063–1070. https://doi.org/10.1021/es203534z.

Graedel, T. E., & Reck, B. K. (2016). Six years of criticality assessments: What have we learned so far? *Journal of Industrial Ecology, 20*, 692–699. https://doi.org/10.1111/jiec.12305.

Grosjean, C., Miranda, P. H., Perrin, M., & Poggi, P. (2012). Assessment of world lithium resources and consequences of their geographic distribution on the expected development of the electric vehicle industry. *Renewable and Sustainable Energy Reviews, 16*, 1735–1744. https://doi.org/10.1016/j.rser.2011.11.023.

Harvey, L. D. D. (2018). Resource implications of alternative strategies for achieving zero greenhouse gas emissions from light-duty vehicles by 2060. *Applied Energy, 212*, 663–679. https://doi.org/10.1016/j.apenergy.2017.11.074.

Huijbregts, M. A. J., Steinmann, Z. J. N., Elshout, P. M. F., Stam, G., Verones, F., Vieira, M. D. M., et al. (2016). *ReCiPe 2016: A harmonized life cycle impact assessment method at midpoint and endpoint level Report I: Characterization.*

Ihle, C. F., & Kracht, W. (2018). The relevance of water recirculation in large scale mineral processing plants with a remote water supply. *Journal of Cleaner Production, 177*, 34–51. https://doi.org/10.1016/j.jclepro.2017.12.219.

McLellan, B. C. (2011). Optimizing location of bulk metallic minerals processing based on greenhouse gas avoidance. *Minerals, 1*. https://doi.org/10.3390/min1010144.

McLellan, B. C. (2014). Streamlining the use of legislated reporting to move to "life of project" sustainability reporting. *International Journal of Mining and Mineral Processing Engineering, 5*. https://doi.org/10.1504/IJMME.2014.058917.

McLellan, B. C., Choi, Y., Ghoreishi-Madiseh, S. A., & Hassani, F. P. (2018). Emissions and the role of renewables: Drivers, potential, projects and projections. In *Mining and sustainable development: Current issues*. https://doi.org/10.4324/9781315121390.

McLellan, B. C., Corder, G. D., & Ali, S. H. (2013). Sustainability of rare earths—An overview of the state of knowledge. *Minerals, 3*. https://doi.org/10.3390/min3030304.

McLellan, B. C., Corder, G. D., Giurco, D. P., & Ishihara, K. N. (2012). Renewable energy in the minerals industry: A review of global potential. *Journal of Cleaner Production, 32*. https://doi.org/10.1016/j.jclepro.2012.03.016.

McLellan, B. C., Diniz Da Costa, J. C., & Dicks, A. L. (2012). Location-specific sustainability metrics: Measuring sustainability space. *Sansai, 6*, 18.

Mudd, G. M. (2007a). An assessment of the sustainability of the mining industry in Australia. *Australian Journal of Multi-Disciplinary Engineering, 5*, 1–12.

Mudd, G. M. (2007b). *The sustainability of mining in Australia: Key production trends and their environmental implications*. Melbourne: Department of Civil Engineering, Monash University and Mineral Policy Institute.

Mudd, G. M. (2010). Global trends and environmental issues in nickel mining: Sulfides versus laterites. *Ore Geology Reviews, 38*, 9–26. https://doi.org/10.1016/j.oregeorev.2010.05.003.

Norgate, T., & Jahanshahi, S. (2010). Low grade ores—Smelt, leach or concentrate? *Minerals Engineering, 23*, 65–73. https://doi.org/10.1016/j.mineng.2009.10.002.

Tokimatsu, K., Höök, M., McLellan, B., Wachtmeister, H., Murakami, S., Yasuoka, R., et al. (2018). Energy modeling approach to the global energy-mineral nexus: Exploring metal requirements and the well-below 2°C target with 100 percent renewable energy. *Applied Energy, 225*. https://doi.org/10.1016/j.apenergy.2018.05.047.

Tuazon, D., & Corder, G. D. (2008). Life cycle assessment of seawater neutralised red mud for treatment of acid mine drainage. *Resources, Conservation and Recycling, 52*, 1307–1314.

Valenta, R. K., Kemp, D., Owen, J. R., Corder, G. D., & Lèbre, É. (2019). Re-thinking complex orebodies: Consequences for the future world supply of copper. *Journal of Cleaner Production, 220*, 816–826. https://doi.org/10.1016/j.jclepro.2019.02.146.

Watari, T., McLellan, B. C., Giurco, D., Dominish, E., Yamasue, E., & Nansai, K. (2019). Total material requirement for the global energy transition to 2050: A focus on transport and electricity. *Resources, Conservation and Recycling, 148*, 91–103.

Watari, T., McLellan, B. C., Ogata, S., & Tezuka, T. (2018). Analysis of potential for critical metal resource constraints in the international energy agency's long-term low-carbon energy scenarios. *Minerals, 8.* https://doi.org/10.3390/min8040156.

Worrall, R., Neil, D., Brereton, D., & Mulligan, D. (2009). Towards a sustainability criteria and indicators framework for legacy mine land. *Journal of Cleaner Production, 17*, 1426–1434. https://doi.org/10.1016/j.jclepro.2009.04.013.

Limits of life cycle assessment in the context of the energy transition and its material basis

8

Fernando Penaherrera[a] and Alexandra Pehlken[b]

[a]Carl von Ossietzky University of Oldenburg, Oldenburg, Germany
[b]OFFIS—Institute for Information Technology, Oldenburg, Germany

Introduction

The development of renewable energy technologies and growing demand for energy is increasing the need for raw materials, both in terms of quantity and variety, for the manufacture of renewable energy components (see Zepf and Wang et al. within this book).

Raw materials are essential for the production of a broad range of products and services, including equipment for renewable energy systems. For example, energy generation using photovoltaic panels requires different quantities of materials such as copper, cadmium, and aluminum. Some raw materials are referred to as "critical materials" (CM) due to their comparatively high economic importance and high risk of supply disruptions (European Commission, 2017, see also Wang et al. and Gilbert within this book). Although geological scarcity is unlikely, as mentioned by Weil et al. in this book, supply risks lie in import dependence, the concentration of production in unstable countries, and the nationalization of mining companies (Buijs, Sievers, Tercero, & Luis, 2012).

The rapid technological innovation cycles and the growth of emerging economies lead to an increasing demand for these highly sought-after metals and minerals. Renewable energy technologies are generally more metals intensive than fossil fuel ones (van der Voet, 2013, see also Wang et al. and Zepf within this book). Consumption of raw material resources implies a decrease in future availability. Resources in this chapter are defined as material goods retrieved from mining efforts, i.e., raw materials.

The term resource efficiency refers to maximizing the output of a system while minimizing its resource consumption, and it is a key element in sustainable

The Material Basis of Energy Transitions. https://doi.org/10.1016/B978-0-12-819534-5.00008-8

development. Methods are required to ensure the ecological use of resources (Klinglmair, Sala, & Brandão, 2014). Scientific methods such as life cycle assessment (LCA) make it possible to take into consideration the impacts caused by different stages of the new energy systems' life cycles and evaluate the total environmental impact of the implementation of new energy technologies and the potential savings in greenhouse gases (GHG) (see also MacLellan within this book). These methods use a cradle-to-grave analysis to evaluate the total impacts of these technologies and compare the normalized results to those of the technologies they are replacing.

LCA modeling of processes and products uses databases about raw material mining, product manufacturing, and the physical properties of a system's input and output flows to assess a product's physical and economic impacts. The success of LCA relies on the quality of the data available in the databases and on the inventories of the processes collected by the LCA practitioner. Access to some databases must be paid for, while other databases are made freely available by government agencies. The raw material data on mining and processing found in these databases varies due to different data sources. The data is either submitted by research projects over many years, or is the result of data gathering during industrial applications. Commonly used databases in Europe include ecoinvent (Wernet et al., 2016), GaBi (thinkstep, 2019), and ProBas (UBA, 2015). Of these three, ecoinvent is the most relevant for LCAs. When conducting an LCA, it is generally not possible to obtain all the necessary data through databases, especially if new technologies are being assessed. Therefore, sometimes assumptions need to be made and this involves the inclusion of uncertainties. The reliability of LCA results increases over time if more than one LCA delivers similar results. The individual assessors consider the economic importance of the materials used, the resource stock, and the energy required to obtain an additional unit of raw material. This is evaluated further in this chapter. However, LCA fails to consider the social aspects related to supply risks, future demands, and impacts on concurring technologies, which can lead to depletion. Newly developed social LCA (SLCA) methodologies could address additional social indicators.

This chapter provides an overview of LCA and the best known methods for assessing raw material consumption, and highlights the advantages and limitations of their application. "Life cycle assessment (LCA) and resource consumption indicators" section provides details about LCA and describes the indicators for assessing raw material consumption and their applications. "Indicators and their relationship to the material demands of emerging renewable energy technologies" section discusses the use of these indicators for evaluating the impacts of material use in renewable energy technologies. "Discussion on the use of indicators for evaluating resource saving potentials" section summarizes the discussion and outlines how indicators could potentially be used to evaluate resource saving potentials. "Conclusion" section offers conclusions and outlines the perspectives for the development of methods for evaluating material consumption in renewable energy technologies, as well as the limitations of these methods.

Life cycle assessment (LCA) and resource consumption indicators

Life cycle assessment (LCA) is a structured, comprehensive, and internationally standardized method. It quantifies all relevant emissions and resources consumed during the entire life cycle of a product or service, as well as the associated environmental and health impacts (ILCD, 2010). LCAs make it possible to compare different technologies, scenarios, and supply chain opportunities (Aggar, Banks, & Dietrich, 2012).

LCAs take into account the full life cycle of a product: from the extraction of resources and the production process, through to the usage phase, recycling, and the disposal of the remaining waste (EU JRC, 2010). The ISO 14040 and 14044 standards provide an indispensable framework for LCA (European Commission, 2010; ISO, 2006). This framework leaves the individual practitioner with a range of choices, which can affect the legitimacy of the results of an LCA study (ILCD, 2010). It shows the high variability of the results from source to source when the same technology is evaluated. This is due to the variety of technology setups and the resources used. For example, the amount of CO_2 emissions that result from photovoltaic energy production can range from 20 to 250 g/kWh, depending on the location of the PV installation and where its raw materials were mined and processed (often these stages do not take place in the same country). This variability is also discussed by McLellan within this book.

The ISO standards define the different steps required in an LCA: Goal and Scope Definition, Inventory Analysis, Impact Assessment, and Interpretation (see Fig. 8.1).

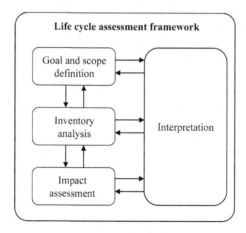

FIG. 8.1

Representation of the different steps of a life cycle assessment.

Credit: Compiled by the author on the basis of ISO (2006). ISO 14064: Environmental management: Life cycle assessment—Principles and framework: ISO.

- *Goal and Scope Definition* involves defining the target product to be analyzed, the reference unit of the product, the system boundary, and the impact methods. Take, for example, an evaluation of the equivalent CO_2 produced by wind energy generation. In this case, the goal is to evaluate the greenhouse gases produced by generating energy through a wind conversion facility. The reference unit can be set as 1 unit of produced energy, 1 kWh, at the grid connection point. The scope can be defined as encompassing the related processes, from raw material gathering, production of wind turbines and related components, transport, installation, location, and commencement of operations. In this example, activities such as decommissioning and recycling of components are excluded from the scope.
- *Inventory Analysis* refers to the collection of data about processes. Mass balances for material flows and thermodynamic relationships (for example, energy per kilogram for energy carriers such as oil) assess the demand for natural resources and the process outputs to the environment. Using the previous example, building models of the system would require information about the quantities of materials and energy necessary to produce the wind turbines, estimations regarding transportation distances, a bill of materials for system parts and components, the expected service life of the components, and estimated energy produced. This provides an inventory of the system inputs and outputs.
- The *Impact Assessment* stage evaluates the impacts of the inputs and outputs of different elementary flows (such as CO_2, methane, and other greenhouse gases) on selected impact categories. Impact categories help us make actionable statements about how emissions influence the environment, and they group different emissions together to express their collective effect on the environment. Impact assessments analyze what are known as "areas of protection," such as human health, the natural environment, issues related to natural resource use, including waste, and many more (EU JRC, 2010). The inputs and outputs are assigned characterization factors for each category. These characterization factors are multipliers that consider the influence of each flow on the selected impact. The results are then normalized to the reference unit, for example, the total equivalent CO_2 emissions produced per unit of energy, or kg CO_2-eq/kWh of electricity produced.
- *Interpretation* refers to the evaluation of the results. This step is used to identify significant issues arising from the results of the Inventory Analysis and Impact Assessment. It focuses on identifying hot spots, a sensitivity analysis, a comparison of technologies, and evaluating the input data and the effects on the results. In the wind energy example, most of the emissions are related to mining, the refining of steel, and the manufacture of glass-reinforced plastic (Wernet et al., 2016).

The LCA practitioner needs to select the most suitable impact assessment methods and indicators for the assessment. Different indicators can be used at different

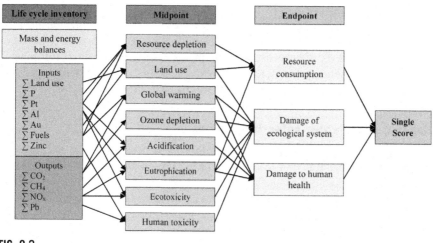

FIG. 8.2

Schematic representation of the impact assessment methods and their aggregation.

Credit: Compiled by the author.

levels of the assessment, since some indicators are developed to generate results about particular impacts on different areas of protection at a middle level (midpoint indicators), while higher-level indicators (endpoint indicators) are used to present a summary of the overall environmental impacts by weighting and aggregating midpoint indicators. Fig. 8.2 provides a graphical representation of this aggregation. As each LCA uses a specific set of categories and indicators, it is difficult and not advisable to compare different LCAs for the same product. Schneider, Berger, and Finkbeiner (2015) and Menoufi (2011) provide a good overview on LCA impact assessment methods and their application to resource consumption.

Indicators are never used individually to produce results. Instead, several indicators are aggregated using the midpoint and endpoint methods. This leads to a joint assessment that reflects all of the life cycle impacts on nature, humans, and the atmosphere. The results of mass balances and life cycle inventories are considered and added to the midpoint indicators, which are then considered and added to the endpoint indicators, whose aggregation results in a single LCA score.

Existing methods for the assessment of resource consumption in LCA relate to the energy and mass of the resource used, exergy or entropy impacts, future consequences of resource extraction (e.g., surplus energy, marginal cost), and diminishing geological deposits. Or they evaluate the environmental damages caused by raw materials extraction (Schneider et al., 2014).

However, we found out that there is a lack of indicators that focus on raw materials. Quantification of mineral resource consumption in LCA has been heavily discussed in the literature of the LCA community (Schneider et al., 2015). There is a multiplicity of approaches for quantifying the effects in the raw materials impact category, as there are several impact category indicators in life cycle impact assessments (LCIA), and there are significant environmental issues associated with the extraction and processing of raw materials (ISO, 2006; Yellishetty, Ranjith, Tharumarajah, & Bhosale, 2009). There is a lack of consensus across impact assessment methods for mineral resource consumption in LCA (Klinglmair et al., 2014), as the different methods were developed using different approaches for the evaluation of impacts.

The role of indicator selection in LCA

Resource indicators in LCA generally combine macro-level territorial resource extraction (such as metal output from mines or the mix of electricity sources used to power mining technologies) and emission inventories (such as CO_2-eq emissions from different energy sources) with the life cycle inventory data of the product under investigation. Indicators are used to assess the impacts of natural resource consumption (Mancini, Sala, Góralczyk, Ardente, & Pennington, 2013, p. 521). Different methods evaluate the impacts of selected processes. The different approaches currently used for the assessment of resource consumption and its impacts are grouped into three categories: resource accounting methods, midpoint methods, and endpoint methods. These methods differ in terms of how they evaluate the influence of each input/output flow, and which area of protection they consider. For example, the Cumulative Energy Demand method is focused exclusively on energy carriers and is used to evaluate the total energy use of a process. In contrast, the CML method evaluates aspects such as the depletion of abiotic resources, the emission of greenhouse gases, and the eutrophication potential. Different methods show different characterizations of the impacts depletion and reveal gaps in the number and types of resources covered (Klinglmair et al., 2014).

Abiotic resource depletion is a suitable measure for focusing on raw materials consumption during the life cycle of a product. Abiotic resources are inorganic materials and include minerals, metals, and fossil fuels. They are the basis for economic growth and play a crucial role in technological developments. They can be used to quantify the depletion (consumption of finite resources) of each distinct material. The abiotic depletion potential is an LCA impact category that quantifies the amount of nonrecoverable resources used (Burchart-Korol & Kruczek, 2016; van Oers, de Koning, Guinée, & Huppes, 2002). The depletion of biotic and abiotic resources is gaining increasing importance for the environmental impact assessments of new technologies (Peters & Weil, 2017, see also Weil et al. within this book). The evaluation of natural resources consumption covers biotic and abiotic resources.

Three approaches are currently used to quantify resource consumption: mass-based accounting (material and energy flow analysis), impact assessments using material flow analysis (MFA) and inventories in LCA, and criticality assessments based on LCIA and an assessment of critical raw materials for the EU economy (Blengini et al., 2017; Mancini, Benini, & Sala, 2015). All three approaches are state of the art within the LCA community and it is up to the practitioner to decide which approach to select. With the mass-based accounting approach, it is possible to determine which life cycle phase has the most impact on the overall life cycle (e.g., most of the CO_2 produced by cars is released into the atmosphere during the use phase). An impact assessment can show which areas of protection are most affected during a product's life cycle (e.g., a product might be carcinogenic to human health). In combination with a criticality assessment, raw materials can be identified that are critical and might become scarce in the future. As Europe relies on raw materials from other continents, the EU has also put a lot of effort into the topic of raw material data and raw material recycling, and it is one of the major research topics within the EU Horizon 2020 research program. As a result, many EU research projects are obliged to feed their project results on materials into the EU's Raw Materials Information System (RMIS) (European Commission, 2019). This platform was developed by the Directorate General Joint Research Centre in cooperation with the Directorate General for Internal Market, Industry, Entrepreneurship, and SMEs (GROWTH).

The next chapter explains how raw material consumption is currently assessed within the LCA framework. The following subsections introduce the different methods that have been developed to assess material consumption and its impacts during the life cycle of a product. Resource accounting methods focus on the calculation of each material's total input and output masses, such as the input of primary copper in kilograms during a product's whole life cycle. As mentioned, midpoint methods weight and aggregate the material flows to calculate environmental impacts, and endpoint methods are used to evaluate overall impacts and draw comparisons with other impact categories.

Resource accounting methods (RAM)

Resource accounting methods are used to measure the material and energy throughputs between natural and anthropogenic systems for the purposes of conducting a material flow analysis (MFA) or developing life cycle inventories (LCI). These focus on calculating the specific mass of materials or energy inputs and outputs of a product (Alvarenga, Oliveira, & de Almeida, 2016; Klinglmair et al., 2014). Resource accounting methods that are suitable for LCA include:

- *Cumulated Material Demand (CMD)* includes all the extracted raw materials in weight units, without differentiation regarding criteria such as scarcity, availability, and exhaustibility (Giegrich, Liebich, Lauwigi, & Reinhardt, 2012). Material criticality is therefore unobserved.
- *Environmental Development of Industrial Products (EDIP)—Nonrenewable Resources* considers the total mass input of distinct natural mineral resources and aggregates them equally regardless of their origin (country or specific deposit) or any scarcity-related parameter. It differentiates between separate raw materials, such as aluminum, silver, and gold (Bigum, Brogaard, & Christensen, 2012).
- *Cumulated Energy Demand (CED)* accounts for resources with energy or heating value such as oil and natural gas, as well as renewable sources such as solar energy and biomass (Alvarenga et al., 2016). This approach takes into consideration

all the energy consumed during the extraction, manufacture, and disposal of a product. It represents a product's total energy demand, calculated as the primary energy consumed during the complete life cycle (Huijbregts et al., 2010; Menoufi, 2011). CED is used when analyzing energy-intensive product systems, such as the manufacture of photovoltaic panels (Wiesen & Wirges, 2017).

The International Reference Life Cycle Data System (ILCD) Handbook published by the European Commission contains further information about these methods (EU JRC, 2010).

Cumulated Energy Demand (CED) and Cumulated Material Demand (CMD) are taken into account in Germany's national sustainability strategy (Giegrich et al., 2012). CED is useful for comparing energy technology products, as it provides information about the energy payback time, or overall efficiency of processes. These two indicators lack information on distinct raw material consumption. The Environmental Development of Industrial Products (EDIP) indicator, on the other hand, presents nonrenewable resource depletion results based on the total mass of each separated resource (Wiesen & Wirges, 2017). However, the EDIP categories for each raw material (such as aluminum) leave aside information such as material origin or production processes.

Midpoint methods

The impact-based methods use life cycle impact assessment (LCIA) methodologies from the life cycle assessment framework. The amount of inventoried material and energy resources is multiplied by characterization factors that represent specific resource-related impacts. These factors are developed based on the properties being studied, such as resource stock depletion, or the economic impacts of depletion, or damage to the environment (Alvarenga et al., 2016). These methods include:

- *CML (developed by the Institute of Environmental Sciences of the University of Leiden)* focuses on environmental impact categories expressed as resource use or emissions released into the environment (Mancini, Benini, & Sala, 2018; van Oers et al., 2002).

For resource depletion, it contains the *Abiotic Depletion Potential* (ADP), which is multiplied by the extracted amount of a given resource and compared with the depletion of 1 kg of antimony (Sb) as a reference (Klinglmair et al., 2014; van Oers et al., 2002). The ADP characterization factors for minerals and fossil fuels are calculated by comparing the resource's extraction rate to its ultimate available reserve (Burchart-Korol & Kruczek, 2016; Schneider et al., 2015; Schödwell & Zarnekow, 2017). The characterization factors are developed using data on ultimate stock reserves, or recoverable reserves, with the ultimate reserves being the baseline. Ultimate stock reserves refer to the quantity of resources that is ultimately available, which is estimated by multiplying the average natural concentration of the resources in the Earth's crust by the mass of the crust (Sala, Benini, Castellani, Vidal-Legaz, & Pant, 2016). The separation of minerals and fossil fuels into distinct categories provides a better representation of the different reserves (Sala et al., 2016; van Oers & Guinée, 2016). The ILCD Handbook (2010) adopted a version of the Abiotic Depletion Potential that is calculated using the reserve base instead of the ultimate reserve estimations.

The *Global Warming Potential (GWP),* also in the CML method, represents a system's equivalent carbon dioxide (CO_2) emissions that contribute to anthropogenic climate change. To quantify the effect of different greenhouse gases and express them as CO_2-equivalent emissions, the emission quantities are multiplied by characterization factors (Myhre et al., 2013). Although not a method for evaluating resource consumption, GWP is the most well-known LCA indicator, and thus worth including for communication purposes.

- The *Criticality Weighted Abiotic Depletion Potential (CWADP)* is a proposed set of indicators that merge the concepts of ADP and criticality. The ADP characterization factors are modified using the criticality values of raw materials (European Commission, 2017). This method provides an overview of the total equivalent criticality of the material content of a product (Koch, Peñaherrera, & Pehlken, 2019).
- The *Eco-Indicator 99 midpoint method* was developed considering the decrease in resource quality based on current extraction rates. The metal depletion category considers all mineral resources as of equal importance and omits substitution possibilities. It uses a lognormal distribution of concentrations of mineral resources to quantify extracted amounts against grade and uses this relationship to calculate marginal effects of present extractions. The concept of surplus energy allows quantifying the required increase of efforts, expressed in MJ (Mega Joule) of energy (Goedkoop & Spriensma, 2001).
- The *ReCiPe midpoint method* was developed with characterization factors for fossil fuels, metals, and minerals, and a different approach is used for each resource. The Mineral Resource Depletion (MRD) impact category was developed using data about deposits and is used to evaluate the depletion of metals and minerals. This method also takes into account decreases in the concentration of minerals found in ores due to their extraction, normalized to kilograms of iron (Alvarenga et al., 2016). This is achieved by using the monetization of surplus energy demand to characterize future efforts for resource extraction. Marginal increases in a resource's extraction cost per kilogram form the basis of the model (Goedkoop, Heijungs, de Schryver, Struijs, & van Zelm, 2009; Klinglmair et al., 2014).

Endpoint methods

Endpoint methods are based on midpoint methods. They are developed to consider the overall environmental burdens associated with resource extraction and use, rather than the most immediate impacts such as the diminishing of stocks or deposits (Alvarenga et al., 2016; Klinglmair et al., 2014). The impact assessment calculations use the different impact categories of the corresponding midpoint methods. The results are assigned to protected areas, weighted in a common unit (e.g., points), and aggregated in order to present the total impacts in a single equivalent score (Schödwell & Zarnekow, 2017). This shows the relative importance of resource depletion in comparison to the other categories, as well as the total environmental impacts of a process.

The calculation of single-point Eco-Indicator scores is intended to aid day-to-day decision-making, and also serves as a general-purpose impact assessment method in LCA (EU JRC, 2010). Endpoint methods are based on various midpoint indicator methods and include:

- The *Eco-Indicator 99 endpoint* evaluates the total impact of using resources by considering the increased workload involved in the extraction of more inaccessible reserves. Categories include damage to ecosystem quality, damage to human health, total resource consumption, and total impacts (Alvarenga et al., 2016; Goedkoop & Spriensma, 2001; Schödwell & Zarnekow, 2017).
- The *ReCiPe endpoint* differentiates between fossil depletion and mineral depletion in the resource category (Alvarenga et al., 2016; European Commission, 2010; Goedkoop, Heijungs, Huijbregts, et al., 2009).

What is the right indicator or method?

The answer to the question "what is the right indicator or method?" is "there isn't one." There are always a variety of indicators and the method is selected on the basis of the overall question under investigation. Table 8.1 presents a summary of the presented methods related to raw material usage during the life cycle. The EU recommends using Cumulated Energy Demand to evaluate the impacts related to total resource consumption and identify potential direct savings in primary resource use, as well as ways to improve process efficiency (European Commission, 2010). EDIP is used to evaluate total resource consumption for each distinct metal. For the midpoint assessment, the European Commission (2010) recommends that the Abiotic Depletion Potential be presented together with GWP for company reports and other communication (Giegrich et al., 2012). The ReCiPe endpoint method is the recommended endpoint LCIA method, as it differentiates between fossil and metal depletion in the resource category (Alvarenga et al., 2016).

Abiotic resource depletion is one of the most debated impact categories. There is no scientifically accurate method for deriving the weighting factors that are used to calculate abiotic depletion based on the input and output flows. Different ways to characterize these weighting factors do exist, such as a decrease in the resource itself, a decrease in international reserves of useful energy, or an incremental change in the environmental impact of extraction processes at some point in the future. Therefore, an assessment of only one indicator can provide insufficient information, which could lead to incorrect conclusions (Drielsma et al., 2016; van Oers & Guinée, 2016). Additionally, the development of different methods has resulted in different characterization factors for distinct materials, and some materials were not taken into account during the development of the methods. For example, neodymium, palladium, and silver are not included in the Eco-Indicator 99 midpoint and endpoint methods for calculating the depletion of metals (Fig. 8.3).

Table 8.1 Methods for assessing the impacts of material resource consumption.

Category	Method	Indicator	Unit	Sources
Resource Account Methods	CED	Primary Energy Resources	MJ_{eq}	Schödwell and Zarnekow (2017)
	CMD	Cumulated Material Demand	kg	Giegrich et al. (2012)
	EDIP2003	Material Depletion	kg	Hauschild and Potting (2005)
Midpoint	ADP GWP	Abiotic Depletion Potential Global Warming Potential	kg_{Sb-eq} kg_{CO2-eq}	Schneider et al. (2015) Guinée (2006), van Oers and Guinée (2016)
	CWADP	Criticality Weighted Abiotic Depletion Potential (Supply Risk and Economic Importance)	kg_{Sb-SR} kg_{Sb-EI}	European Commission (2017), Koch et al. (2019)
	ME	Eco-Indicator 99— Mineral Extraction	points	Goedkoop and Spriensma (2001)
	MRD	ReCiPe—Mineral Resource Depletion	kg_{Fe-eq}	Klinglmair et al. (2014), Goedkoop, Heijungs, de Schryver, et al. (2009)
Endpoint	Eco-Indicator 99	Resources—Total	points	Goedkoop and Spriensma (2001)
	ReCiPe	Total Resource Depletion	points	Goedkoop, Heijungs, de Schryver, et al. (2009)

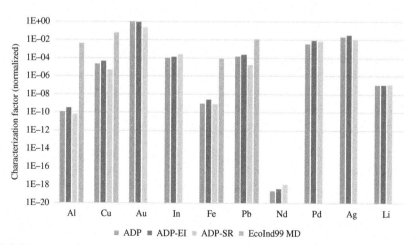

FIG. 8.3

Comparative characterization factors for different impact assessment methods.

Credit: Compiled by the author.

Fig. 8.3 shows the characterization factors of different impact assessment methods. The figures are normalized in logarithmic scale in relation to the material with the highest characterization factor. Some metals are absent in specific methods, since they were not considered when the methods were developed (e.g., indium is not included in the Eco-Indicator 99 method).

Indicators and their relationship to the material demands of emerging renewable energy technologies

Increased usage of renewable energy technologies has raised awareness among different stakeholders that this development also relates to increasing demand for critical metals. In most cases, metals are categorized as critical depending on their economic relevance and what specific technologies they are used for. Renewable energy technologies contain critical materials, but they provide a lot of renewable energy for society and this is neglected in the assessments. If policies that reduce vulnerabilities are implemented (e.g., improving relations with China, thereby safeguarding the supply of metals), then metals might be perceived as more critical in the future (Månberger & Stenqvist, 2018). Criticality assessments might be a suitable method for policy makers to regulate the market.

Although data is often sparse, the available information about the material demands of some renewable energy technologies suggests that current practices are likely to lead to scarcity for some metals in the not too distant future (Graedel & Erdmann, 2012). Thus, raw material productivity could be increased by minimizing material inputs and reusing production waste, which would result in lower environmental impacts and less consumption of scarce resources, despite the same energy output (Giegrich et al., 2012). Currently, the methods for impact assessment in LCA recommended by the ILCD Handbook (European Commission, 2010) cover some scarcity-related issues, such as depletion from mines.

For example, LCA can assess the environmental implications of the mining and processing of raw materials, including critical raw materials. Based on such assessments, indicators can be identified that support the definition of materials that have high environmental impacts. Similarly, life cycle data for critical raw materials can provide valuable insights into the options for managing these materials at the end of their life stage, particularly when evaluated using a material flow analysis (Mancini, Sala, et al., 2015).

LCIA makes it possible to address the overall impact of the implementation of these technologies and to investigate what consequences the future depletion of these metals will have on specific impact categories, such as resource consumption. The existing impact assessment methods lack measures for assessing future environmental impacts, since the methods were developed with information related to current (albeit outdated) natural reserves. The consequences of further exploitation of these metals need to be analyzed. Furthermore, the materials considered "critical" vary according to market conditions. Because the supply of metals and minerals is largely met by mining and is tied to particular geological formations that are generally un-evenly distributed around the globe, some countries and regions will remain dependent on foreign resources (Buijs et al., 2012). This supply dependence may change if the anthropogenic stocks are considered (see Goddin, Chapter 12 within this book).

For example, the EDIP2003 and CML2002 methodologies are mass-based, yet show considerable discrepancies in their mineral depletion indicators related to iron (Klinglmair et al., 2014). The CED method only addresses energy requirements and is good for developing indicators like Energy Payback Time, but it makes no distinction between raw material usage. These energy-based indicators also do not consider or provide information about material sources. EDIP2003 is useful for generating information about the total use of separated resources, but the source and quality of these resources is not considered when aggregating the mass flows for the indicators.

LCA approaches to resource depletion have mostly focused on the geophysical availability of a given mineral or metal. The impact assessment methods presented do not consider the constraints of environmental legislation or the political, economic, and geostrategic considerations in the producing countries. These latter aspects are not covered by the methods discussed, which were developed based on information about reserves and extraction rates. Criticality indicators are calculated for minerals and metals, so that economic and geopolitical factors can be included in LCA. However these criticality indicators are not used when researchers evaluate the material demands of products (Klinglmair et al., 2014). Short-term geological aspects are considered in social life cycle assessments (SLCA), where the methods used include categories such as the supply risk of products based on vulnerabilities to supply disruption within the supply chain (Cimprich, Karim, & Young, 2018; Gemechu, Helbig, Sonnemann, Thorenz, & Tuma, 2016).

Due to the dependence on raw materials and their availability in new technologies, recent literature proposed the integration of resource criticality assessments in the life cycle sustainability assessment (LCSA) framework (Mancini et al., 2018, p. 727; Koch et al., 2019). Criticality assessments attempt to include concepts of criticality in the existing methods for evaluating resource depletion. Other methods such as Eco-Indicator 99 and ReCiPe focus on the increased effort involved in the marginal production of certain materials. These methods focus on current reserves and extraction rates, and do not consider future developments. The ecological scarcity method weights flows based on "critical amounts" of elementary flows, which are set according to political and strategic goals.

Discussion on the use of indicators for evaluating resource saving potentials

Energy production is a key priority in sustainability assessments. Energy-related materials are thus essential to sustainable development (Green, Espinal, Traversa, & Amis, 2012), and the depletion of abiotic and biotic resources is a fundamental issue for sustainability assessments (Klinglmair et al., 2014).

Due to the crucial role of metals in society, now and in the future, it is necessary to safeguard their supply and understand how to minimize the environmental impacts associated with their production and use. Altogether, the primary production of metals presently accounts for 7%–8% of total global energy consumption. This figure may increase because of the processing of lower grade ores, which requires more energy. Yet the energy efficiency of production processes can be, and is being, increased substantially (van der Voet, 2013).

Several measures have been proposed to minimize the impacts of resource consumption and reduce material depletion: recycling, substitution, efficiency, and longer product life spans (Buijs et al., 2012, see also chapter by Goddin within this book). Ensuring the supply of raw materials for industry requires: (a) expansion and improved efficiency of ore mining and metal extraction, (b) substitutions at the level of materials and technologies, (c) resource efficiency in production and use, and (d) recycling, ensured by recyclable designs, recirculation strategies, and efficient recycling technologies (Marscheider-Weidemann et al., 2016). Raw materials savings and efficiency should be equally reflected by the indicator selection in LCA.

As we have shown, there is no ideal indicator or method available today in LCA to assess impacts caused by raw materials and criticality. Mass and unit-based metrics provide insufficient information about the benefits of the recycling rates of critical materials. Complete cradle-to-grave LCA analyses make it possible to evaluate the total benefit in terms of reductions in cumulative energy demand and material depletion. Proposed indicators such as Avoided Environmental Burden (AEB) evaluate these potential benefits (Oswald, 2013). The secondary production of metals is much less energy intensive than primary production, with energy savings ranging from 55% (lead) to 98% (palladium or aluminum). An obvious way to reduce energy use related to metals production is therefore to increase secondary production (van der Voet, 2013, p. 94). However, recycling of dissipated metals is restricted by related energy demands and costs (Graedel & Erdmann, 2012).

Data or assessments data sets of comparable quality are not always available for all raw materials included in the life cycle analysis, as seen in indicators such as the Abiotic Depletion Potential. A common database for the comparison of different raw materials would increase the quality of information on relevant issues (e.g., the influence of emerging uses, recycling potential, the influence of resource nationalism on supply trends, cradle-to-grave environmental impacts) (Buijs et al., 2012). Today, there is still great uncertainty when comparing different LCIA related to raw material and criticality impacts, but there are already projects underway that take this issue seriously into account, such as the "Sustainable Management of Critical Raw Materials" project (EIT, 2017).

Conclusion

Renewable energy technologies generate electricity and heat from renewable sources such as wind and sunlight. But manufacturing such technologies requires nonrenewable raw materials, which should ideally be used in sustainable ways. Indicator-based assessments can be useful in this regard and should be taken into consideration when working out how to best manage nonrenewable resources (van der Voet, 2013). Most of the commonly used approaches in LCA are still incapable of predicting the physical scarcity of raw materials in the future and the consequences for sustainable material use. Given the crucial role played by metals in society, now and in the future, it is necessary to safeguard their supply and understand how to minimize the environmental impacts associated with their production and use.

As resource efficiency is considered a key element for sustainable development, there is an increasing need for suitable methods to address the sustainability of resource use (Klinglmair et al., 2014). LCA can assess the efficiency of a process by considering the whole life cycle, using methods such as cumulative energy demand, or by measuring savings in abiotic depletion potential. Using only one method for the assessment of resource consumption does not provide sufficient information.

Resource criticality has so far received more attention outside the LCA community and is gaining importance in policy making (European Commission, 2014). It is therefore desirable to use a resource depletion indicator that reflects the supply criticality of a given resource, subject to economic, political, and strategic influences, in addition to mere availability in the natural environment (Klinglmair et al., 2014). Indicators such as the criticality weighted abiotic depletion potential represent initial attempts to merge the concepts of criticality with resource depletion (Koch et al., 2019). There is a lack of fully dynamic criticality analysis, although some authors have conducted static assessments of different time periods, or analyzed stocks and flows of materials over time (Du & Graedel, 2011; Graedel et al., 2012). Thus, new approaches are required to incorporate the dynamic aspect of criticality (Roelich et al., 2014).

While a mass-based approach relies on consistent data-gathering systems, the LCA approach has the advantage of translating physical flows into multiple metrics expressing some societal concerns, such as climate change or damage to human health. This makes it possible to consider a multiplicity of perspectives when making decisions (Mancini, Benini, et al., 2015).

LCIA endpoint methods aggregate resource consumption into one cumulative indicator. This does not take into account the different uses of resources and their origin (Brentrup, Küsters, Lammel, & Kuhlmann, 2002). Resource availability assessments need to go beyond the currently established evaluation in LCA, and shift from assessing availability as an environmental impact to defining resource availability as a sustainability problem (Schneider, 2014).

A long-term integrated assessment of the use of the full range of resources (including critical ones) could aid the development of schemes for minimizing resource usage during the energy transition (Viebahn et al., 2015). In addition to the assessment of primary resource availability, future studies need to consider differentiating between primary and secondary resources (Schneider, 2014).

The lack of available information makes it difficult to assess the long-term sustainability of the implementation of renewable energy technologies and their material requirements. For example, the number of data sets for primary metals is generally much higher than those available for recycled materials. There are not enough reliable data sets for recycled metals. Assessment of this data will enable the inclusion of these materials into environmental impact analyses. For some recycled metals, there is no data available in LCA databases about the recovery of metals from renewable energy technologies.

Most methods acknowledge the depletion of natural resources from a functional point of view. This neglects the intrinsic value of minerals (van Oers et al., 2002). Expanding the analysis of the impacts of material usage requires an evaluation of sociopolitical, economic, and environmental dimensions. These need to be developed in addition to the existing analysis of LCA and social LCA. Changes in economic data caused by fluctuating demand, exploration and supply cycles, as well as politics and socioeconomic trends make the inclusion of a temporal dimension inescapable (Drielsma et al., 2016; Sala et al., 2016).

Addressing dependencies between indicators can also show the effects of reducing material use in other impact categories. Studies on the impacts of the effects of renewable energy technologies generally omit effects such as "rebound" or "leapfrogging," which could disrupt the supply of raw materials to manufacturers. Including a dynamic analysis in these methodologies could provide information for policy development.

These indicators could be further developed so that they also consider the benefit to society. Indicators such as energy productivity and material productivity could link the depletion of natural resources with a system's total gains in economic productivity.

There is still a lot of work to be done in relation to sustainable management in the context of life cycle assessments and renewable energy. There is great demand for materials and it is constantly increasing.

References

Aggar, M., Banks, M., & Dietrich, J. (2012). *Data centre life cycle assessment guidelines.* White Paper #45. Edited by The Green Grid. The Green Grid. Beaverton, OR, USA https://www.on365.co.uk/files/3714/3203/1600/Data-Centre-Life-Cycle-Assessment-Guidelines.pdf. (Accessed 29 June 2018).

Alvarenga, R., Oliveira, I., & de Almeida, J. (2016). Evaluation of abiotic resource LCIA methods. *Resources, 5*(13). https://doi.org/10.3390/resources5010013.

Bigum, M., Brogaard, L., & Christensen, T. (2012). Metal recovery from high-grade WEEE. A life cycle assessment. *Journal of Hazardous Materials, 207–208,* 8–14. https://doi.org/10.1016/j.jhazmat.2011.10.001.

Blengini, G. A., Blagoeva, D., Dewulf, J., Torres de Matos, C., Nita, V., Vidal-Legaz, B., et al. (2017). *Assessment of the methodology for establishing the EU list of critical raw materials.* Background Report Luxemburg: Edited by Publications Office of the European Union. European Commission.

Brentrup, F., Küsters, J., Lammel, J., & Kuhlmann, H. (2002). Impact assessment of abiotic resource consumption conceptual considerations. *International Journal of Life Cycle Assessment, 7*(5), 301–307. https://doi.org/10.1007/BF02978892.

Buijs, B., Sievers, H., Tercero, E., & Luis, A. (2012). Limits to the critical raw materials approach. *Proceedings of the Institution of Civil Engineers—Waste and Resource Management, 165*(4), 201–208. https://doi.org/10.1680/warm.12.00010.

Burchart-Korol, D., & Kruczek, M. (2016). Depletion of abiotic resources in the steel production in Poland. *Metalurgija, 55*(3), 531–534.

Cimprich, A., Karim, K. S., & Young, S. B. (2018). Extending the geopolitical supply risk method: material "substitutability" indicators applied to electric vehicles and dental X-ray equipment. *International Journal of Life Cycle Assessment, 23*(10), 2024–2042. https://doi.org/10.1007/s11367-017-1418-4.

Drielsma, J. A., Russell-Vaccari, A. J., Drnek, T., Brady, T., Weihed, P., Mistry, M., et al. (2016). Mineral resources in life cycle impact assessment—defining the path forward. *International Journal of Life Cycle Assessment, 21*(1), 85–105. https://doi.org/10.1007/s11367-015-0991-7.

Du, X., & Graedel, T. E. (2011). Global in-use stocks of the rare Earth elements: a first estimate. *Environmental Science & Technology, 45*(9), 4096–4101. https://doi.org/10.1021/es102836s.

EIT. (2017). *Sustainable Management of Critical Raw Materials.* European Institute of Innovation and Technology (EIT) https://suscritmat.eu/. (Accessed 22 November 2019).

EU JRC. (2010). *ILCD handbook—Background document: Analysis of existing environmental impact assessment methodologies for use in life cycle assessment* (1st ed.). Ispra, Italy https://eplca.jrc.ec.europa.eu/uploads/ILCD-Handbook-LCIA-Background-analysis-online-12March2010.pdf. (Accessed 21 August 2019).

European Commission. (2010). *ILCD handbook—General guide on LCA— Detailed guidance.* Luxembourg, Luxembourg: European Commission. http://publications.jrc.ec.europa.eu/repository/bitstream/JRC48157/ilcd_handbook-general_guide_for_lca-detailed_guidance_12march2010_isbn_fin.pdf. (Accessed 3 August 2018).

European Commission. (2014). *Review of the list of critical raw materials for the EU and the implementation of the raw materials initiative.* Brussels, Belgium https://eur-lex.europa.eu/legal-content/EN/TXT/PDF/?uri=CELEX:52014DC0297&from=EN. (Accessed 18 July 2018).

European Commission. (2017). *Study on the review of the list of critical raw materials.* Criticality assessments.

European Commission. (2019). *Raw materials information system.* https://rmis.jrc.ec.europa.eu/. (Accessed 4 November 2019).

Gemechu, E. D., Helbig, C., Sonnemann, G., Thorenz, A., & Tuma, A. (2016). Import-based indicator for the geopolitical supply risk of raw materials in life cycle sustainability assessments. *Journal of Industrial Ecology, 20*(1), 154–165. https://doi.org/10.1111/jiec.12279.

Giegrich, J., Liebich, A., Lauwigi, C., & Reinhardt, J. (2012). *Indikatoren/Kennzahlen für den Rohstoffverbrauch im Rahmen der Nachhaltigkeitsdiskussion.* Dessau-Roßlau, Germany: Umweltbundesamt. http://www.uba.de/uba-info-medien/4237.html. (Accessed 21 August 2018).

Goedkoop, M., Heijungs, R., de Schryver, A., Struijs, J., & van Zelm, R. (2009). *ReCiPe 2008. A life cycle impact assessment method which comprises harmonised category indicators at the midpoint and the endpoint level. Report I: Characterisation.* Amersfoot, The Netherlands https://www.leidenuniv.nl/cml/ssp/publications/recipe_characterisation.pdf. (Accessed 17 October 2019).

Goedkoop, M., Heijungs, R., Huijbregts, M., de Schryver, A., Struijs, J., & van Zelm, R. (2009). *ReCiPe 2008—A life cycle impact assessment method which comprises harmonised category indicators at the midpoint and the endpoint level. Report I: Charaterisation* (1st ed.). The Hague, the Netherlands: Edited by Ministry of Housing, Spatial Planning and the Environment (VROM). Ministry of Housing, Spatial Planning and the Environment (VROM).

Goedkoop, M., & Spriensma, R. (2001). *The eco-indicator 99. A damage oriented method for life cycle impact assessment. methodology report.* Amersfoort, the Netherlands https://www.pre-sustainability.com/download/EI99_methodology_v3.pdf. (Accessed 17 October 2019).

Graedel, T. E., Barr, R., Chandler, C., Chase, T., Choi, J., Christoffersen, L., et al. (2012). Methodology of metal criticality determination. *Environmental Science & Technology, 46*(2), 1063–1070. https://doi.org/10.1021/es203534z.

Graedel, T. E., & Erdmann, L. (2012). Will metal scarcity impede routine industrial use? *MRS Bulletin, 37*(4), 325–331. https://doi.org/10.1557/mrs.2012.34.

Green, M. L., Espinal, L., Traversa, E., & Amis, E. J. (2012). Materials for sustainable development. *MRS Bulletin, 37*(4), 303–309. https://doi.org/10.1557/mrs.2012.51.

Guinée, J. B. (2006). *Handbook on life cycle assessment. Operational guide to the ISO standards. vol. 7.* Dordrecht: Kluwer Academic Publishers (Eco-efficiency in Industry and Science).

Hauschild, M., & Potting, J. (2005). *Spatial differentiation in life cycle impact assessment—The EDIP2003 methodology.* Environmental News 80 https://www2.mst.dk/udgiv/publications/2005/87-7614-579-4/pdf/87-7614-580-8.pdf. (Accessed 3 August 2018).

Huijbregts, M. A. J., Hellweg, S., Frischknecht, R., Hendriks, H. W. M., Hungerbühler, K., & Hendriks, J. A. (2010). Cumulative energy demand as predictor for the environmental burden of commodity production. *Environmental Science & Technology, 44*(6), 2189–2196. https://doi.org/10.1021/es902870s.

ILCD. (2010). *ILCD handbook—General guide on LCA—Detailed guidance.* https://eplca.jrc.ec.europa.eu/uploads/ILCD-Handbook-General-guide-for-LCA-DETAILED-GUIDANCE-12March2010-ISBN-fin-v1.0-EN.pdf. (Accessed 22 August 2019).

ISO. (2006). *ISO 14064: Environmental management: Life cycle assessment—Principles and framework: ISO.*

Klinglmair, M., Sala, S., & Brandão, M. (2014). Assessing resource depletion in LCA. A review of methods and methodological issues. *International Journal of Life Cycle Assessment, 19*(3), 580–592. https://doi.org/10.1007/s11367-013-0650-9.

Koch, B., Peñaherrera, F., & Pehlken, A. (2019). *Criticality and LCA—Building comparison values to show the impact of criticality on LCA.* In *ICSD 2019: 7th International Conference on Sustainable Development.*

Månberger, A., & Stenqvist, B. (2018). Global metal flows in the renewable energy transition: Exploring the effects of substitutes, technological mix and development. *Energy Policy, 119*, 226–241. https://doi.org/10.1016/j.enpol.2018.04.056.

Mancini, L., Benini, L., & Sala, S. (2015). Resource footprint of Europe. Complementarity of material flow analysis and life cycle assessment for policy support. *Environmental Science & Policy, 54*, 367–376. https://doi.org/10.1016/j.envsci.2015.07.025.

Mancini, L., Benini, L., & Sala, S. (2018). Characterization of raw materials based on supply risk indicators for Europe. *International Journal of Life Cycle Assessment, 23*(3), 726–738. https://doi.org/10.1007/s11367-016-1137-2.

Mancini, L., Sala, S., Góralczyk, M., Ardente, F., & Pennington, D. (2013). Life cycle assessment and criticality of raw materials: Relationship and potential synergies. In *6th international conference on life cycle management. The 6th international conference on life cycle management, vol. 2013. Gothenburg, Sweden* (pp. 520–523).

Mancini, L., Sala, S., Recchioni, M., Benini, L., Goralczyk, M., & Pennington, D. (2015). Potential of life cycle assessment for supporting the management of critical raw materials. *International Journal of Life Cycle Assessment, 20*(1), 100–116. https://doi.org/10.1007/s11367-014-0808-0.

Marscheider-Weidemann, F., Langkau, S., Hummen, T., Erdmann, L., Tercero Espinoza, L., Angerer, G., et al. (2016). *Summary—Raw materials for emerging technologies 2016.* Berlin, Germany https://www.isi.fraunhofer.de/content/dam/isi/dokumente/ccn/2016/Zukunftstechnologien_Zusammenfassung_EN.pdf. (Accessed 26 August 2019).

Menoufi, K. (2011). *Life cycle analysis and life cycle impact assessment methodologies: A state of the art.* Master's Thesis.

Myhre, G., Shindell, D., Bréon, F. M., Collins, W., Fuglestvedt, J., Huang, J., et al. (2013). Anthropogenic and natural radiative forcing. Climate change 2013: The physical science basis. In *Contribution of working group I to the fifth assessment report of the intergovernmental panel on climate change.*

Oswald, I. (2013). *Environmental metrics for WEEE collection and recycling programs* [Dissertation zur Erlangung des Doktorgrades (Dr. rer. nat.) der Mathematisch-Naturwissenschaftlichen Fakultät der Universität Augsburg]. Doctoral Thesis Augsburg, Germany: University of Augsburg. https://opus.bibliothek.uni-augsburg.de/opus4/frontdoor/deliver/index/docId/2312/file/Dissertation_Irina_Oswald.pdf. (Accessed 25 July 2018).

Peters, J., & Weil, M. (2017). Providing a common base for life cycle assessments of Li-ion batteries. *Journal of Cleaner Production, 171,* 704–713. https://doi.org/10.1016/j.jclepro.2017.10.016.

Roelich, K., Dawson, D. A., Purnell, P., Knoeri, C., Revell, R., Busch, J., et al. (2014). Assessing the dynamic material criticality of infrastructure transitions: A case of low carbon electricity. *Applied Energy, 123,* 378–386. https://doi.org/10.1016/j.apenergy.2014.01.052.

Sala, S., Benini, L., Castellani, V., Vidal-Legaz, B., & Pant, R. (2016). *Environmental footprint—Update of life cycle impact assessment methods, DRAFT for TAB (status: May 2, 2016).* https://ec.europa.eu/environment/eussd/smgp/pdf/JRC_DRAFT_EFLCIA_resources_water_landuse.pdf. (Accessed 22 August 2019).

Schneider, L. (2014). *A comprehensive approach to model abiotic resource provision capability in the context of sustainable development* [Doktor der Ingenieurwissenschaften—Dr.-Ing.—genehmigte Dissertation]. Doctoral Thesis Berlin, Germany: Technical University of Berlin. Faculty III Process Sciences https://depositonce.tu-berlin.de/bitstream/11303/4460/1/schneider_laura.pdf. (Accessed 12 July 2018).

Schneider, L., Berger, M., & Finkbeiner, M. (2015). Abiotic resource depletion in LCA – background and update of the anthropogenic stock extended abiotic depletion potential (AADP) model. *International Journal of Life Cycle Assessment, 20*(5), 709–721. https://doi.org/10.1007/s11367-015-0864-0.

Schneider, L., Berger, M., Schüler-Hainsch, E., Knöfel, S., Ruhland, K., & Mosig, J. (2014). The economic resource scarcity potential (ESP) for evaluating resource use based on life cycle assessment. *International Journal of Life Cycle Assessment, 19*(3), 601–610. https://doi.org/10.1007/s11367-013-0666-1.

Schödwell, B., & Zarnekow, R. (2017). *Kennzahlen und Indikatoren für die Beurteilung der Ressourceneffizienz von Rechenzentren und Prüfung der praktischen Anwendbarkeit.* Dessau-Roßlau, Germany: Abschlussbericht. Umweltbundesamt. http://www.umweltbundesamt.de/publikationen. (Accessed 13 August 2018).

thinkstep. (2019). *GaBi LCA databases.* thinkstep http://www.gabi-software.com/international/databases/gabi-databases/. (Accessed 4 November 2019).

UBA. (2015). *ProBas. Prozessorientierte Basisdaten für Umweltmanagement-Instrumente.* Umweltbundesamt https://www.probas.umweltbundesamt.de/php/index.php. (Accessed 4 November 2019).

van der Voet, E. (2013). Environmental risks and challenges of anthropogenic metals flows and cycles. In *Nairobi, Kenya, Paris, France: United Nations Environment Programme, International Resource Panel (Report 3 of the Global Metal Flows Working Group of the International Resource Panel of UNEP)*. https://orbit.dtu.dk/files/54666484/Environmental_Challenges_Metals_Full_Report.pdf. (Accessed 21 August 2019).

van Oers, L., de Koning, A., Guinée, J. B., & Huppes, G. (2002). Abiotic resource depletion in LCA. In *Improving Characterisation Factors for Abiotic Resource Depletion as Recommended in the New Dutch LCA Handbook*. https://www.leidenuniv.nl/cml/ssp/projects/lca2/report_abiotic_depletion_web.pdf. (Accessed 22 August 2019).

van Oers, L., & Guinée, J. (2016). The abiotic depletion potential. background, updates, and future. *Resources, 5*(1), 16. https://doi.org/10.3390/resources5010016.

Viebahn, P., Soukup, O., Samadi, S., Teubler, J., Wiesen, K., & Ritthoff, M. (2015). Assessing the need for critical minerals to shift the German energy system towards a high proportion of renewables. *Renewable and Sustainable Energy Reviews, 49*, 655–671. https://doi.org/10.1016/j.rser.2015.04.070.

Wernet, G., Bauer, C., Steubing, B., Reinhard, J., Moreno-Ruiz, E., & Weidema, B. (2016). The ecoinvent database version 3 (part I). overview and methodology. *International Journal of Life Cycle Assessment, 21*(9), 1218–1230. https://doi.org/10.1007/s11367-016-1087-8.

Wiesen, K., & Wirges, M. (2017). From cumulated energy demand to cumulated raw material demand. The material footprint as a sum parameter in life cycle assessment. *Energy, Sustainability and Society, 7*(1), 25. https://doi.org/10.1186/s13705-017-0115-2.

Yellishetty, M., Ranjith, P. G., Tharumarajah, A., & Bhosale, S. V. (2009). Life cycle assessment in the minerals and metals sector. A critical review of selected issues and challenges. *The International Journal of Life Cycle Assessment, 14*(3), 257. https://doi.org/10.1007/s11367-009-0060-1.

Critical resources, sustainability, and future generations

Björn Koch

COAST—Centre for Environment and Sustainability, Carl von Ossietzky University of Oldenburg,
Oldenburg, Germany

Introduction

The energy transition from fossil fuels toward a greater use of renewable energy within the energy system is seen as a major prerequisite for reaching the goals defined by the Paris climate agreement. New technologies play an important role in this endeavor, together with energy conservation and improved energy efficiency. The increasing demand for advanced technologies on a large and global scale goes hand in hand with the shift toward a new material basis (see Wang et al., Zepf, Weil et al. in this book). Many of the elements required to manufacture renewable energy technologies are defined as critical raw materials ("critical materials"), and some are even defined as "conflict resources" (see Prause in this book).

The concept of "critical resources" uses criticality assessments to determine the criticality status of materials. These criticality assessments are always based on a frame of reference which itself is highly "context-specific [and] subject to the perception of how elements may be affected by supply risk and their importance" (Hayes & McCullough, 2018). This context is affected by geopolitical and technological circumstances, which are "influencing the frames of reference that designate criticality" (Hayes & McCullough, 2018). Criticality assessments are generally purely economically driven and focus on the supply risks and the economic importance of the materials (see Zepf, Gilbert, Wang et al. in this book). Criticality assessments rarely take social, environmental, or ethical considerations and impacts into account.

The concept of "conflict resources" emerged in the late 1990s. To invoke international action, a definition of this concept was proposed by Global Witness, an international NGO founded in 1993 that campaigns to end environmental and human rights abuses driven by the exploitation of natural resources and corruption in the global political and economic system (GW, 2019). Global Witness states that "conflict resources are natural resources whose systematic exploitation and trade in a context of conflict contribute to, benefit from, or result in the commission of serious violations of human rights, violations of international humanitarian law or violations amounting to crimes under international law" (GW, 2010). This definition is solely based on human rights and international law, and in contrast to the critical resources' concept, it does not take into account any economic or ecological considerations. If we assume that international law is "more ethical than legal in nature," as proposed by legal theorist John Austin (Scott, 1905), and human rights are "certain moral guarantees" (Fagan, 2019), then the concept of conflict resources is purely based on and driven by moral and ethical considerations.

This chapter takes a closer look at the concepts of "critical resources" and "conflict resources" in the context of sustainability and sustainable development. Based on this, the moral and ethical aspects of these concepts are examined, as well as some general problems that arise in relation to these aspects.

Critical resources, conflict resources, and sustainability

The concepts of "critical resources" and "conflict resources" are based on very different basic values: economic stability and growth on the one hand, and universal ethical and moral principles on the other. Sometimes it seems impossible to take these basic values into account simultaneously, and the strange situation arises in which gold and tin are among the four most prominent conflict minerals—tantalum, tin, tungsten, and gold (known as the 3TGs)—yet are not considered critical resources by some criticality assessments. For instance, the European Commission's criticality assessment currently rates both gold and tin as "not critical"[a] (EC, 2017).

Furthermore, neither concept takes into account basic and general considerations regarding sustainability, because of their exclusive and very strong emphasis on their respective basic values.

A common and widely used concept of sustainability is the classic three-pillar model. Within this model, the "economy" pillar, the "society" pillar, and the "environment" pillar form the foundation for sustainability. These three pillars are interdependent, and sustainability can only be achieved if all three pillars are respected and balanced.

With a primary focus on universal ethical and moral principles, the concept of conflict resources seems to be independent of the three pillars of sustainability, while the concept of critical resources only takes the "economy" pillar into account. The other two pillars—"society" and "environment"—are only considered as side aspects, if at all. At least basic considerations on environmental issues are mentioned in the European Commission's criticality assessment (EC, 2017) in the subchapters "Environmental and regulatory issues." But these environmental assessments tend to

---

[a] Both gold and tin have been classified as candidate materials and have consequently been subjected to the European Commission's criticality assessment since 2014. But in the 2014 and 2017 reports they were identified as "not critical."

focus on the European Union's REACH regulation, which has an economic focus: "REACH is a regulation of the European Union, adopted to improve the protection of human health and the environment from the risks that can be posed by chemicals, while enhancing the competitiveness of the EU chemicals industry" (ECHA, 2007). Social factors are only mentioned at one point in the European Commission's criticality assessment (EC, 2017, p. 249): "…what may become problematic is the ability to access these resources as a result of social, environmental, political, and economic factors." This reduces social factors solely to the possibility of being a "potential problem" for the supply chain. Social aspects are not dealt with within a sustainability perspective, which would consider the social impact of the extraction and processing of critical resources.

But there is a deeper problem regarding the classification of these concepts within the concept of sustainability: the synonymous use of the terms "sustainability" and "sustainable development."

There are many definitions of "sustainability" and "sustainable development" currently available. Nevertheless, these terms have often been used ambiguously since sustainability rose to prominence and gained global attention a few decades ago. Even though "there exists an obvious semantic difference, and implicit focus in meaning, this distinction is not always present in the literature, especially in reference to the pillars formulation" (Purvis, Mao, & Robinson, 2019).

The very basic meaning of sustainability, which itself is not unique, is "the ability to exist constantly" (Wikipedia, 2019). In the context in which the term "sustainability" is currently used, it refers to the constant existence of individual people, civilizations, and humankind. It is questionable whether this anthropocentric view does justice to the matter, or if a more biocentric or even holistic approach should be taken. But either way, the existence of a biosphere that enables the ongoing existence of human beings is a crucial prerequisite, regardless how this biosphere is defined, and whether its value is defined anthropocentrically or intrinsically. This results in an implicational relationship between the biosphere and the "constant existence" of individual people, civilizations, and humankind, with the biosphere being a necessary condition for this "constant existence," and thus for sustainability.

The most used and cited definition of sustainable development can be found in a report published by the United Nations World Commission on Environment and Development in 1987, titled "Our Common Future" and commonly called the Brundtland Report: "Humanity has the ability to make development sustainable to ensure that it meets the needs of the present without compromising the ability of future generations to meet their own needs. The concept of sustainable development does imply limits—not absolute limits but limitations imposed by the present state of technology and social organization on environmental resources and by the ability of the biosphere to absorb the effects of human activities. But technology and social organization can be both managed and improved to make way for a new era of economic growth" (WCED, 1987). It highlights the relationship between all three pillars of sustainability: social, environmental, and economic.[b] A generally accepted understanding of the three-pillar model of sustainability is that these three pillars are interdependent, and that sustainability can only be achieved if all three pillars are respected and balanced. It is the task of sustainable development to balance these pillars, and when a balanced state is achieved this is referred to as "sustainable." The more unbalanced these pillars are, the less sustainable the (sustainable) development will be. But what is happening in the definition above is that the environmental and social pillars are being instrumentalized, and become "manageable tools" that need to be improved in the name of economic growth. Obviously, a much stronger emphasis is placed on the economic pillar. Thus, this pillar is overvalued and it outweighs the others. This imbalance causes a set of (what I call) "tilted pillars of sustainability." These tilted pillars lead to an unbalanced system, and are an indicator of unsustainable development.

Altogether, this results in a strange and paradoxical situation in which the most commonly used definition of sustainability actually defines sustainable development instead, and this definition itself tends to lead to unsustainable development. So it is highly debatable whether we will ever be able to reach the end point of sustainability, or even have the chance to get close to it, if we keep on following the wrong interpretation and using this counterproductive definition.

Finally, there is another challenge that arises from the synonymous use of the terms "sustainability" and "sustainable development," which underlines the noninterchangeable nature of these terms. Sustainability is like the state of a system and a goal to aspire to, while sustainable development is a process that should enable us to reach and maintain this state of sustainability. The pillars of sustainability form the interdependent structure that needs to be kept in balance, in order for the process of sustainable development to actually be sustainable and contribute toward a state of sustainability. The Brundtland Report sets this straight at the end of its definition of sustainable development: "Yet in the end, sustainable development is not a fixed state of harmony, but rather a process of change in which the exploitation of resources, the direction of investments, the orientation of technological development, and institutional change are made consistent with future as well as present needs" (WCED, 1987). A very good explanation of the relationship between these two noninterchangeable terms is provided by Shaker: "The term 'sustainability' should be viewed as humanity's target goal of human-ecosystem equilibrium (homeostasis), while 'sustainable development' refers to the holistic approach and temporal processes that lead us to the end point of sustainability" (Shaker, 2015).

The United Nations (UN, 2015) is addressing this relationship by stipulating 17 Sustainable Development Goals that are to be met by 2030. These sustainable development goals can be seen as an approach that aims to achieve "humanity's target

---

[b] Actually it even introduces a fourth pillar: technology. It is generally questionable whether the classical three pillars are sufficient for portraying the overall system of sustainability or sustainable development. There are many interpretations about the amount, composition, and interdependencies of the pillars of sustainability. The most frequently added pillars are culture, technology, and governance.

goal of human-ecosystem equilibrium." However, all 17 goals must be pursued and achieved in order to create the conditions for a state of sustainability.

Critical resources, conflict resources, and future generations

While the previous section concentrated on the concepts of "critical resources" and "conflict resources" within sustainability and sustainable development, we now take a closer look at the moral and ethical aspects of these two concepts.

Given that the concept of *conflict resources* is directly derived from human rights and international law, it is purely based on and driven by moral and ethical considerations. It suggests that all aspects of sustainability and sustainable development—economic, social, and environmental—should be subordinated to the overall goal: the freedom and the equality in dignity and rights of all human beings without distinction of any kind (UN, 1948).

The moral and ethical aspects of the concept of *critical resources* are a bit more complex. As we have seen before, this concept is based on criticality assessments that focus mainly or exclusively on economic aspects. Social and environmental aspects are instrumentalized in the name of economic growth within the perspective of sustainable development. This way, any moral and ethical claims related to social and environmental issues are of secondary importance, and are substantially neglected.

Furthermore, this imbalanced focus on economic aspects is passed on to the criticality status of resources through the criticality assessments. Any social and environmental impact caused by the extraction and processing of a critical resource has no influence on the resource's criticality status. Similarly, a resource's impact on the society and the environment would not change just because its criticality status might change from "not critical" to "critical" in the future. If there are any moral and ethical aspects related to the extraction and processing of a resource, then they are not dependent on its criticality status. The same conclusions are reached when we look at the use phase of critical resources in products. The use of a resource has an environmental or social impact that is also independent of the material's criticality status.

These "moral and ethical detachments" are caused by the use of a sustainable development concept that instrumentalizes social and environmental values in the name of economic growth as the foundation in criticality assessments.

If we decide to stick with this concept of sustainable development, all key values are connected to economic aspects. Within criticality assessments, which "should help to implement the 2030 'agenda on sustainable development and its sustainable development goals,' [...] the main parameters used to determine the criticality of the material for the EU are economic importance and supply risk" (EC, 2017). Many experts have published articles on criticality assessments during the last decade and their input will not be further evaluated here (Cimprich et al., 2019; Gemechu, Helbig, Sonnemann, Thorenz, & Tuma, 2016; Graedel et al., 2012). While a resource itself cannot be inherently critical, its criticality *status* is determined by two main parameters: economic importance and supply risk. These parameters are country or region specific, and vary globally as they are affected by geopolitical and technological circumstances that are "influencing the frames of reference that designate criticality" (Hayes & McCullough, 2018).

If we accept this context as the basic link between criticality and sustainable development, we can find a possible starting point where moral and ethical aspects are reattached. This starting point results from the Brundtland Report, which stipulates that sustainable development should meet "the needs of the present without compromising the ability of future generations to meet their own needs" (WCED, 1987). Thus, if a resource's economic importance and supply risk are the nucleus of the criticality definition, which itself is based on the concept of sustainable development, and this sustainable development has to meet the needs of future generations, now and in the future and independent of location, then we have moral and ethical aspects that can be directly linked to a resource's criticality: our moral obligations towards future generations and their needs. These moral obligations are duties that we owe to future generations and their needs. They are duties that result from considerations of "right" and "wrong," and which we are not legally obligated to fulfill.

Supply risk "reflects the risk of a disruption in the EU supply of the material" (EC, 2017). This risk refers to the primary supply. It is influenced by the geopolitical situation in the countries that produce and process raw materials, and can be mitigated by substitution and recycling. The notion of economic importance reflects the importance of a material for the EU economy. In order to take future generations and their future needs into account, these parameters and their underlying values need to be reevaluated by applying them to an extended timeline. By taking this perspective, the global reserves, the overall global resources, and the scarcity of elements become dominant factors of supply risk. An element's supply risk will also be influenced by future energy technologies, which will have a direct impact on its economic importance (see Weil et al. in this book). These factors—global reserves, global resources, and future technologies—are heavily influenced by the way the present generation chooses to act and satisfy its needs.

One of the basic questions that arises from the issues discussed above is: Do we have moral obligations towards future generations at all? Many essays have been written about this topic, especially since sustainability and environmental ethics have attracted great attention over the past decades, but there is still no clear answer to this question. So in this chapter I am going take the position outlined in the Brundtland Report, and the point of view of common sense, and assume that we do have moral obligations towards future generations. The next section looks at some of the general problems related to these obligations with a focus on critical resources.

Parfit's paradox and future generations

One of the general problems related to our moral obligations towards future human beings—as individual members of future generations—is the paradoxical situation that arises if future human beings would start to complain about the lack of action taken by previous generations to ensure that their right to be able to meet their own needs is not violated. This paradox, which is called Parfit's paradox (Saugstad, 2003), is often brought up in the debate about moral obligations towards future generations as an argument to deny or at least weaken obligations we might have towards future human beings. It is based on the assumption that every single decision we take has an impact on the way the future unfolds, and so with every decision taken there is a change in the passage of time. If, in the future, human beings complain about the lack of action taken in the past that resulted in the violation of their right to be able to meet their own needs, then these human beings will base their complaints on the assumption that the decisions taken by the past generations led to their current situation and circumstances. The paradox arises because on the one hand, these human beings might have been able to meet their own needs if past generations had made other decisions, but on the other hand, those decisions would have changed the passage of time, and these human beings might never have been born. In that hypothetical case, apart from the fact that past decisions and the passage of time cannot be undone, it makes no sense to complain about "wrong" decisions taken by past generations as these human beings would never have existed. This results in the (purely theoretical, but not really existing) "choice" facing these human beings: they can "choose" between their own existence, and the miserable circumstances in which they live due to an (unavoidable) inability to meet their own needs. Apart from the paradox, at first there does not seem to be any choice in this situation: What could be more valuable than your own existence?

It might be true that we do not have any obligations towards particular human beings who have not been born yet, as they will probably never been born. But we know that there are going to be future human beings, and based on the definition of sustainable development in the Brundtland Report, we have moral obligations towards them although we do not know exactly who they are going to be.

But we can imagine circumstances in which a person might not be too sure if their continued existence would be the better choice. These circumstances would involve an unbearable amount of suffering and struggles, which would make their life not worth living. A life where all one's very basic needs are barely satisfied. A life where this person would regret being born. A life without dignity. At a global level, however, these circumstances reflect an extreme postapocalyptic dystopia. A dystopian scenario that is sometimes conjured up in conjunction with exaggerated, extreme prophecies about the effects of climate change, and which has been the subject in many cli-fi books and movies like *Waterworld*, *The Day After Tomorrow*, *Wall-E*, and *Interstellar* (to name just a few well-known blockbusters).

So does this mean we can continue to exploit the planet and its resources as long as we do not leave it in a dystopian state? Obviously not, because this would imply that we only have to ensure that future generations are able to meet the minimum and basic needs for survival, and not "their own needs" as outlined by the Brundtland Report. Furthermore, would this kind of life, in which people are only able to meet their minimum needs for survival, provide the dignity demanded in the Universal Declaration of Human Rights (UN, 1948, Article 1)?

Apparently there is more involved in our moral obligations towards future generations than the bare minimum, that is, the continued existence of humanity. This is why the definition of sustainable development in the Brundtland Report specifies "the ability of future generations to meet their own needs" (WCED, 1987) and not "their minimum needs for survival."

The needs of future generations

The interpretation of *their own needs* is a crucial element in our moral obligations towards future generations. As we have seen, these needs exceed humanity's minimum requirements for survival. A biosphere that enables the ongoing existence of human beings is a crucial prerequisite for these needs. And finally, assuming that the Universal Declaration of Human Rights will remain universal in the future, "the economic, social, and cultural rights indispensable for his dignity and the free development of his personality" (UN, 1948, Article 22) will also be part of *their own needs*. Thus, this approach equates the definition of *their own needs* with the definition of *our own needs*. The only specific differences between "our needs" and "their needs" will be those that result from future changes in the economic, social, and cultural systems, as well as the related rights.

Technology has always influenced these three systems, but the inclusion of information technology in our daily lives has dramatically changed all three systems within a very short period of time. Nobody could have foreseen the dramatic rise of social media, and its massive impact on our social lives, 25 or 30 years ago. This impact has dramatically changed the needs related to our social and cultural systems within less than one generation.[c] How could past generations who lived 100 or more years ago have had any idea about the present generation's social needs, and the social rights associated with them? The same is true for economic and cultural needs and rights. Assuming that technological progress will not slow down in the future, and that this progress will have dramatic impacts on the future of our economic, social, and cultural systems, it seems doubtful that we will be able to reasonably or even approximately predict the needs of future generations. The chances are that this progress and its impacts will exceed our wildest hypotheses. What's more, the more temporally distant the future generation under consideration, the more impossible it will be to make any realistic forecast.

Given that we cannot know for sure what the needs of future generations will look like, does this mean we can simply continue exploiting the planet and its resources? Obviously not. However, the situation is not entirely clear-cut. The definition of sustainable development in the Brundtland Report explicitly stresses that "the needs of the present [...] generation" (WCED, 1987) should be met as well. So on the one hand, sustainable development must meet the needs of the present generation, which can be interpreted as justification for the further exploitation of the planet. But on the other hand, the needs of the present generation must be met "without compromising

---

[c] Assuming that a generation, as an average period of time, is roughly 30 years.

the ability of future generations to meet their own needs" (WCED, 1987). Apparently these two provisions can be contradictory, which leaves us with an ethical dilemma: Does the present generation have to sacrifice some of its needs to enable future generations to meet their needs? Or do future generations need to give up some of their needs, so that the present generation is able to meet its needs? And if so, how many needs must be sacrificed by which generation?

How many future generations do we need to consider?

The task of determining the amount of needs that a particular generation must sacrifice exposes another fundamental problem related to the moral obligations we have towards future generations: How many generations do we need to include in our calculations? From a sustainability point of view, which aims to ensure the ongoing existence of individual people, civilizations, and humankind, we will have to include all future generations. But if we have to take all future generations into consideration, we will end up with an extremely large and unpredictable number of human beings. Then, if we want to determine the amount of a nonrenewable resource, like oil, that every present and future human being is entitled to, we will end up with an extremely small amount of oil—if any at all—that can be consumed per person. Ultimately, this would lead to the paradoxical situation that basically no present or future human being would be able to use any oil at all. This conclusion would have an extreme impact on the present and probably a few succeeding generations, as our daily lives still depend heavily on oil and oil-based products. It is doubtful whether this scenario would still allow the present generation to meet its own needs. Altogether, the involvement of all future generations is not only impossible to implement on a practical level, as we will never know the exact number of human beings to consider, but it is also contradictory to the concepts of sustainable development and human rights. This problem could be solved by reducing the number of future generations that need to be considered. Of course uncertainties would remain, but at least we would get some results to start working with. Unfortunately, such a reduction would also generate another unsolvable problem: How many future generations or years do we have to take into consideration? We will end up having to decide where we want to make the cut—and no matter what reason we choose to back up our decision, it will most likely be an arbitrary and therefore an unfair and unjust one. In the end, there is nothing left but the techno-optimistic hope, that the impact of future technologies will supersede the impact of our unfair decision before the time of the cut is reached. Otherwise, it will be hard to explain this decision to the generations following the cut. No matter what reasons are presented to them, they will feel that they have been wronged.

Taking precautions for future generations

On the one hand, we are stuck with the dilemma that the actual consumption and depletion of resources involved in meeting the present generation's needs is very likely to compromise the ability of future generations to meet their own needs. Currently, there does not seem to be a way to solve this problem without cutting back at least some of the present generation's needs. But on the other hand, we cannot predict the needs of future generations. Furthermore, in order to make any kind of estimate, we would have to make an unfair and unjust decision about how many generations or years will be taken into consideration. This makes it impossible to reasonably determine which of the present generation's needs must be curtailed, or forfeited altogether, in order to achieve the necessary sustainable development. Unfortunately, this dilemma can be daunting or even paralyzing, as there are currently no convincing or promising, and especially no convenient, solutions available. There is no "correct" way out, and no "just for all generations" solution, no matter which decision is taken. But decisions and actions need to be taken and fortunately there is the principle of precaution, which "enables decision-makers to adopt precautionary measures when scientific evidence about an environmental or human health hazard is uncertain and the stakes are high" (EP, 2015). In *The Imperative of Responsibility*, Hans Jonas introduced a precautionary principle based on the "priority of the bad over the good prognosis" (Jonas, 1979). In his view, the advances in technological development and their ever more unpredictable and long-lasting effects have enabled us to harm not only the existing people, but also future generations and all humankind. For Jonas, the existence of humanity as a "cosmic responsibility" is at stake, and because the stakes are so high, he prefers to prioritize "the bad over the good prognosis" (Jonas, 1979).

Nowadays, the principle of precaution is widely accepted. It has even been included into the Rio Declaration on Environment and Development as the principle of choice for protecting the environment: "In order to protect the environment, the precautionary approach shall be widely applied…" (UNCED, 1992, Principle 15). In short, the precautionary principle suggests that policymakers and lawmakers have an inherent responsibility to protect the public from exposure to harm whenever a plausible risk has been identified by scientific investigation. Given the dilemma described above, the best approach for dealing with this situation would arguably be to extend this principle to future generations and their needs.

Conclusions

The increasing demand for advanced technologies caused by the energy transition from fossil fuels toward renewable energy has massively changed the material basis of the energy system. These advanced technologies are considered a major driver of cleaner economic growth that will help to safeguard sustainable development in the future. Many resources included in the new material basis are classified as critical resources, because of their economic importance and supply risk. Some are classified as conflict resources, because their extraction and trade are associated with serious violations of human rights and/or international humanitarian law.

The concept of conflict resources is purely based on and driven by moral and ethical considerations. It is directly derived from human rights and international law, and implies moral obligations towards all present and future human beings. The concept of critical resources, on the other hand, is based on criticality assessments that define the criticality status of a resource. The economic importance and the supply risk are the most common sets of parameters used in criticality assessments. These assessments are used to facilitate future economic growth and are therefore based on the concept of sustainable development. In the general sense, and the commonly used definition in the Brundtland Report, sustainable development has to meet "the needs of the present without compromising the ability of future generations to meet their own needs" (WCED, 1987). These "needs of future generations" imply that each present generation has moral obligations towards future generations. It is virtually impossible to determine such moral obligations as we do not know the composition of the future generations, the total number of people, or what their needs will be. Taking all future generations into account would mean that the present generation is essentially unable to use any resources at all, because the distributed amount of resources that each generation could use would be infinitely small. A cutoff point would need to be determined in order to limit the amount of people considered, but this point would most likely be arbitrary, and therefore represent an unfair and unjust decision. Furthermore, the present generation's current consumption and depletion of resources to meet its needs is very likely to compromise the ability of future generations to meet their own needs. It is highly unlikely that all these needs can be met without cutbacks, which leaves us in a situation with great uncertainties and presumably contradictory premises.

In a situation like this, the best approach might be to follow the principle of precaution, and extend it to future generations and their needs.

References

Cimprich, A., Bach, V., Helbig, C., Thorenz, A., Schrijvers, D., Sonnemann, G., et al. (2019). Raw material criticality assessment as a complement to environmental life cycle assessment: Examining methods for product-level supply risk assessment. *Journal of Industrial Ecology*, 23, 1226–1236. https://doi.org/10.1111/jiec.12865.

European Chemicals Agency (ECHA). (2007). *Understanding REACH-ECHA*. https://echa.europa.eu/regulations/reach/understanding-reach. (Accessed November 5, 2019).

European Commission (EC). (2017). *Communication from the Commission to the European Parliament, the Council, the European Economic and Social Committee and the Committee of the regions on the 2017 list of critical raw materials for the EU*. Brussels: European Commission. Available from https://ec.europa.eu/growth/sectors/raw-materials/specific-interest/critical_en.

European Parliament (EP). (2015). *The precautionary principle: Definitions, applications and governance*. Think Tank. https://www.europarl.europa.eu/thinktank/en/document.html?reference=EPRS_IDA(2015)573876. (Accessed November 12, 2019).

Fagan, A. (2019). Human rights. In *The Internet encyclopedia of philosophy*. ISSN 2161-0002 https://www.iep.utm.edu/hum-rts/. (Accessed November 4, 2019).

Gemechu, E. D., Helbig, C., Sonnemann, G., Thorenz, A., & Tuma, A. (2016). Import-based indicator for the geopolitical supply risk of raw materials in life cycle sustainability assessments. *Journal of Industrial Ecology*, 20, 154–165. https://doi.org/10.1111/jiec.12279.

Global Witness (GW). (2010). *Definition of conflict resources—Natural resources in conflict.* Archived from the original on June 2, 2010. https://web.archive.org/web/20100602192048 http://www.globalwitness.org/pages/en/definition_of_conflict_resources.html. (Accessed 4 November 2019).

Global Witness (GW). (2019). *Exposing human rights abuse & global corruption.* Global Witness. https://www.globalwitness.org/en/about-us/. (Accessed December 2, 2019).

Graedel, T. E., Barr, R., Chandler, C., Chase, T., Choi, J., Christoffersen, L., et al. (2012). Methodology of metal criticality determination. *Environmental Science & Technology, 46*(2), 1063–1070. https://doi.org/10.1021/es203534z.

Hayes, S. M., & McCullough, E. A. (2018). Critical minerals: A review of elemental trends in comprehensive criticality studies. *Resources Policy, 59*, 192–199. https://doi.org/10.1016/j.resourpol.2018.06.015.

Jonas, H. (1979). *Das Prinzip Verantwortung. Versuch einer Ethik für die technologische Zivilisation.* Frankfurt: Suhrkamp.

Purvis, B., Mao, Y., & Robinson, D. (2019). Three pillars of sustainability: In search of conceptual origins. *Sustainability Science, 14*(3), 681–695. https://doi.org/10.1007/s11625-018-0627-5.

Saugstad, J. (2003). Future generations of people [lecture]. In *Moral responsibility towards future generations of people–Utilitarian and Kantian ethics compared.* http://folk.uio.no/jenssa/Future%20Generations.htm. (Accessed 16 October 2019).

Scott, J. B. (1905). The legal nature of international law. *Columbia Law Review, 5*(2), 124–152. Available from https://www.jstor.org/stable/1109809.

Shaker, R. R. (2015). The spatial distribution of development in Europe and its underlying sustainability correlations. *Applied Geography. 63*, 304–314. https://doi.org/10.1016/j.apgeog.2015.07.009.

UN. (1948). *The universal declaration of human rights.* United Nations. https://www.un.org/en/universal-declaration-human-rights/index.html. (Accessed 4 November 2019).

UN. (2015). *The sustainable development agenda—United Nations Sustainable Development.* https://www.un.org/sustainabledevelopment/development-agenda/. (Accessed 12 November 2019).

UNCED. (1992). *Rio declaration on environment and development.* https://www.un.org/en/development/desa/population/migration/generalassembly/docs/globalcompact/A_CONF.151_26_Vol.I_Declaration.pdf. (Accessed 2 January 2020).

Wikipedia. (2019). *Sustainability.* https://en.wikipedia.org/wiki/Sustainability. (Accessed 4 November 2019).

World Commission on Environment and Development (WCED). (1987). *Our common future.* Available from https://sustainabledevelopment.un.org/content/documents/5987our-common-future.pdf.

Conflicts related to resources: The case of cobalt mining in the Democratic Republic of Congo

Louisa Prause

Albrecht Daniel Thaer-Institute of Agricultural and Horticultural Sciences, Agricultural and Food Policy Group, Humbolt University of Berlin, Berlin, Germany

Introduction

The development and increased production of new, environmentally friendly technologies in the context of energy transitions is strongly dependent on raw materials such as minerals and metals. The German Raw Materials Agency estimates that by 2035 four times the current production of lithium, three times that of heavy rare earths and one and a half times that of light rare earths and tantalum will be required for the production of 42 key future technologies for the energy transition (Marscheider-Weidemann, Langkau, Hummen, Erdmann, & Tercero Espinoza, 2016). Such estimates are not unique to Germany, but characteristic for the rising demand of metals for the energy transition more generally.

A key component of the energy transition is the electrification of the transport sector. National action plans in several European countries envision a sharp increase in electric vehicles in the near future to reduce the CO_2 emissions produced by the transport sector. In 2016, the transport sector was responsible for 24% of energy-based CO_2 emissions worldwide (IEA, 2018). Today, many countries in Europe are aiming for electric vehicles (see Sonnberger, Gilbert, Weil et al. within this book). This political strategy for greening the transport sector is an attempt to reconcile Western modes of production, consumption, and mobility with climate protection. Individual motoring is not questioned as such, but is supposed to be made compatible with climate protection through e-mobility.

To date, lithium-ion batteries are the most promising technology for the electrification of the transport sector, due to their high voltages, low weight, and high energy densities that enable greater driving range (Ziemann, Grunwald, Schebek, Müller, & Weil, 2013). The expansion of the production of lithium-ion batteries has led to a

The Material Basis of Energy Transitions. https://doi.org/10.1016/B978-0-12-819534-5.00010-6

sharp increase in demand for several raw materials necessary for their production. Since this increase in demand could so far not be addressed through recycling, it has led to the expansion of mining frontiers, many of which are located in the Global South. The expansion of mining is often highly contested and frequently sparks conflicts (Bebbington & Bury, 2013; Conde & Le Billon, 2017; Dietz & Engels, 2017). UNEP estimates that about 40% of global conflicts in the past 60 years were linked to the extraction of resources (UNEP, 2009). These conflicts are diverse and range from social conflicts over compensation payments and conflicts over dispossession and resettlements, through to violent conflicts and civil wars. However, conflicts around the extraction and use of raw materials for energy transitions not only take place at the site of extraction but also along the global supply chains. Nongovernmental organizations (NGOs) and social movements in the Global North are increasingly advocating for due diligence regulations, in order to guarantee the human rights of those involved in the supply chains of green technologies. Conflicts around the extraction of minerals are thus not merely local phenomena. They take place at different points of the global production network of green technologies and involve a diverse range of actors.

In this chapter, I analyze the conflicts related to the production of cobalt for use in lithium-ion batteries at different points of the global production network. Cobalt is a key raw material required for lithium-ion batteries. A lithium-ion battery for an average electric car contains 4.5–9 kg of cobalt. Tesla's large electric cars need up to 15 kg of cobalt (Frankel, 2016). Cobalt mining is concentrated in the Democratic Republic of Congo (DRC) and the industry is highly contested. I will show how the increase in demand for cobalt has changed mining in the DRC and caused different types of conflicts at various points of the global production network of lithium-ion batteries. The aim of this contribution is twofold: First, I wish to highlight the transregional interdependencies between energy transitions in the Global North and conflicts over the production of raw materials in the Global South. Second, I aim to reveal how the global production network of lithium-ion batteries connects conflicts over the expansion of industrial mining in the Global South to struggles over the use and sourcing of raw materials for energy transitions in the Global North.

I thereby bring together the debates about energy transitions and mining-related conflicts. Raw materials have been addressed in the debate on energy transitions with regard to supply chains (e.g., Olivetti, Ceder, Gaustad, & Fu, 2017), geopolitical challenges (e.g., Scholten, 2018), and the secure provision of strategic metals for environmentally friendly technologies, which are known as critical materials (e.g., Gunn, 2014; Wellmer et al., 2019). So far, only very few contributions have looked at the relationship between energy transitions and conflicts related to mineral extraction (Anlauf, 2016; Church & Crawford, 2018; Prause & Dietz, N.D.). Mining conflicts have been discussed at length in a different strand of the literature, mainly in the context of the mineral price boom of the early and mid-2000s (for an overview see Conde, 2017). However, contributions to this debate focus on the causes of and multiple dimensions of conflicts (e.g., Bebbington & Bury, 2013; Conde & Le Billon, 2017; Dietz & Engels, 2017) and rarely link their findings back to the energy transition. This is also true for conflicts around the extraction of cobalt (e.g., Katz-Lavigne, 2019; Sovacool, 2019).

Methodologically my analysis is based on an extensive literature review, reports by international NGOs, and newspaper articles. I also build upon my own experiences and observations gathered from February 2019 to September 2019 as an employee for the Berlin-based NGO PowerShift e.V., which is engaged in issues such as due diligence regulations related to the extraction and processing of minerals and metals.

The chapter is structured as follows: In section two, I develop a theoretical framework based on political ecology, an action-oriented understanding of conflict, and the concept of Global Production Networks (GPN).[a] In the third section, I show how the expansion of electric mobility changes the demand for cobalt and sketch a few key nodes of the GPN, relating cobalt to the production of electric vehicles. I then identify different types of conflict related to the mining of cobalt in the DRC, before turning to conflicts around the extraction and supply of cobalt in the Global North. In the sixth section I analyze how cobalt mining conflicts at different points of the global production network are related to one another. Finally, in the conclusions, I summarize my key findings and reflect upon the potential links and contradictions between struggles in the Global South and North related to the use of raw materials for green technologies.

Analyzing mining conflicts from a political-ecological perspective

In line with works from the field of political ecology, I understand mining conflicts as both distributional and cultural conflicts. One the one hand, there are struggles over the distribution of access to, use of, and control over raw materials and the distribution of negative impacts (Martinez-Alier, 2009). Local communities struggle against issues like displacement, the destruction of their livelihoods or ecosystems through mining, and for an appropriate share of the profits from resource extraction. On the other hand, communities also struggle for the recognition of their relationship to nature and the cultural meanings they attach to nature and places (Escobar, 2006). This becomes visible when protesters demand different development models or the implementation of a reciprocal relationship between society and nature. In order to grasp these phenomena empirically, I rely on the definition provided by Bonacker and Imbusch (2006), who define conflicts as interactive, social action between at least two individual or collective actors with interests, goals, or needs that they perceive as contradictory. Conflicts are structured by power relations and the diverging interests of the actors involved, which in turn are impacted by the social conditions of a given society (Dietz & Engels, 2014). I understand power as the ability of actors to control their own relationships, as well as those of others, with natural resources such as land, water, metals, and minerals. Thus, power means being able to determine who has access to and who appropriates natural resources, how and under which institutional conditions this is to be done, and for what purposes (Bryant & Bailey, 1997).

---

[a] The theoretical framework of this chapter is a revised and expanded version of the theoretical framework developed together with Kristina Dietz in Prause and Dietz (forthcoming). I am grateful to Kristina Dietz for her valuable comments and thoughts on this section.

This definition of conflicts places the actions and demands of the social actors involved—as well as the social power relations—at the center of the analysis of mining conflicts. However, to analyze power relations in conflicts over the extraction of resources, it is necessary to look beyond the specific place of extraction. We need a transnational analytical perspective in order to grasp the links between the sites of extraction and sites where minerals and metals are processed, for example, the industrial sites where cobalt is processed, and where battery cells and car bodies are produced. I therefore combine my approach to mining conflicts with the concept of global production networks (GPN). Global production networks are "the globally organized nexus of interconnected functions and operations by firms and non-firm institutions through which goods and services are produced and distributed" (Coe, Hess, Yeung, Dicken, & Henderson, 2004: p. 471). The concept emphasizes the social and political embeddedness of global supply chains and includes non-firm actors such as nation states and labor in the analysis (Coe, Dicken, & Hess, 2008). Bridge (2008) stresses that GPNs in the extractive sector differ from other sectors due to the "landed" nature of assets. Minerals and metals are located in very specific places. Unlike other commodities, for example, textiles, the location of their extraction and production is not arbitrary, but has to take place where deposits are found. The materiality of minerals and metals thus exerts a strong influence on production networks and extractive industries are strongly embedded in the institutional arrangements of the states hosting mineral deposits (Bridge, 2008: p. 413). Recently, labor geography has turned to the GPN approach and reconceptualized GPNs as "networks of embodied labor" (Cumbers, Nativel, & Routledge, 2008: p. 372). Contributions from this perspective focus on how social relations of production, class conflict, and resistance shape GPNs and the power relations within them (e.g., Pye, 2017). While labor has been placed prominently within the GPN debate, this has not yet been the case for social movements and civil society actors. However, as I will show in this chapter, these actors actively shape how resources are extracted and global production networks are organized. This combination of the GPN-approach and an understanding of conflicts around mining that is rooted in political ecology allows me to analyze why conflicts emerge at different points in the global production network and how they relate to one another.

Cobalt: From the DRC to the electric car

The analysis of the cobalt supply chain for the production of electric vehicles will focus on two points that have sparked the most visible conflicts so far: the upstream end of cobalt extraction in the DRC and the downstream end of production and consumption of electric vehicles in the Global North. The expansion of e-mobility has contributed significantly to the rising demand for cobalt. In 1990, only 1% of global cobalt production was used to manufacture batteries; in 2015, 49% went into battery production (Haus, 2017). Meanwhile, global cobalt production has almost doubled in the past 10 years from 75,900 tons in 2008 to 140,000 tons in 2018 (US Geological Survey, 2009, 2019A) (Fig. 10.1).

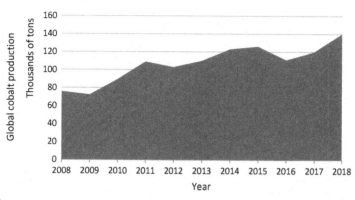

FIG. 10.1

Global cobalt production 2008–2018.

Credit: Compiled by the author on the basis of US Geological Survey data.

The demand for cobalt will most likely increase further in the next 5 years. The US Geological Survey (April 18, 2019) estimates that global demand for cobalt will increase to around 222,000 tons by 2025, of which more than half, around 117,000 tons, will be used to produce batteries. About half of the globally known cobalt reserves (approx. 3.4 million tons) are located in the Democratic Republic of Congo, which currently produces about 60% of the world's cobalt (US Geological Survey, 2019B). Cobalt mining in the DRC is concentrated in the southern regions of Lualaba and Haut-Katanga (Fig. 10.2).

Cobalt is mined industrially as a by-product of copper mining. However, it is estimated that about 15%–20% of the cobalt mined in the DRC stems from artisanal mining and that about 110,000–150,000 artisanal miners are active in cobalt mining (Al Barazi et al., 2017; Amnesty International, 2016). Upstream, the global production network of industrially mined cobalt from the DRC is dominated by two large-scale mining companies. The Swiss company Glencore that operates the mines Mutanda and Katanga, as well as the Chinese company China Molybdenum that operates the Tenke Fungurume Mine (TFM). Until 2016 TFM was owned by Freeport-McMoRan Copper & Gold Inc. and Lundin Mining. In 2018, the output of these mines accounted for almost half of all cobalt production in the DRC (Sanderson, 2019).

Both artisanally and industrially mined cobalt from the DRC is primarily exported to China, while a smaller portion is exported to Finland. China has been the world's largest producer of refined cobalt since 2004 (Al Barazi et al., 2017). The biggest cobalt refining company is China-based Zhejiang Huayou Cobalt Company Ltd (Huayou), which also processes cobalt mined in the DRC (Amnesty International, 2016). Outside of China, Glencore and Freeport-McMoRan are important producers of refined cobalt (Al Barazi et al., 2017). China exports much of its refined cobalt to South Korea, as that country plays an important role in the manufacturing of cobalt products. However, in 2015 China also occupied the leading position in the

FIG. 10.2

Provinces of the DRC.

Credit: CC/Moyogo, highlights added by the author, original map available here: https://de.wikipedia.org/wiki/
Datei:Prov-Congo-Kinshasa_-_2006.fr.svg#filehistory

manufacturing stage, followed by South Korea and Japan (Sun, Hao, Liu, Zhao, & Song, 2019). In 2015, the United States and China accounted for 22% and 20.7% of cobalt consumption, respectively (Sun et al., 2019). Nearly 80% of China's cobalt consumption in 2015 and 2016 can be attributed to the rechargeable battery industry (Olivetti et al., 2017). It is difficult to precisely trace the cobalt mined in the DRC to consumer end products. However, a study conducted by Amnesty International suggests that cobalt from the DRC that is refined by Huayou is then used by battery component manufacturers such as Ningbo Shansan, BAK, Tianjin Lishen and Samsung SDI, and the resulting components are deployed in cars produced by Volkswagen and Daimler, as well as numerous other well-known consumer goods including electronics made by Samsung and Apple (Amnesty International, 2016).

Conflicts around cobalt mining: The upstream end in the DRC

Cobalt has been mined in the DRC since the 1920s, first by Belgian-owned *Union Minière du Haute-Katanga* (UMHK), and after independence by the state-owned mining company *Gécamines* (see Gilbert within this book). In the years following the DRC's independence in 1960, *Gécamines* gained almost monopolistic control over the mining sector, which it sustained until the Congo wars of 1996–2003 (Faber, Krause, & Sanchez de la Sierra, 2017). During this period the state enterprise almost completely collapsed and industrial copper-cobalt mining in Lualaba and Haut-Katanga came to a standstill. From that point on, artisanal mining expanded in the region (Geenen & Cuvelier, 2019). Toward the end of the Congo wars the World Bank increased its pressure on the DRC to liberalize the mining sector and in 2002 the government passed a new investor-friendly mining law. In this process, *Gécamines* was forced to make several of its mines available for foreign investors, which resulted in a renewed expansion of industrial cobalt-copper mining in the DRC (Rubbers, 2019).

The recent expansion of industrial copper-cobalt mining was highly contested and sparked numerous conflicts in the DRC. One key type of conflict emerged between artisanal miners and industrial mining companies. Most of the artisanal cobalt mining activities in the DRC are done informally. Artisanal miners mostly operate without a legal permit and often on industrial concessions. Confrontations and open conflicts frequently emerged when industrial mining companies, with the support of state security forces, attempt to remove artisanal miners from their concessions (De Koning, 2009). These conflicts were particularly intense in the period from 2005 to 2008, when foreign mining companies took over many of the concessions formally owned by *Gécamines*, and industrial cobalt-copper mining recommenced and expanded in the DRC (Hönke, 2010; Ngoie & Omeje, 2008; Rubbers, 2019). Violent confrontations repeatedly took place between artisanal miners on the one hand and state and private security forces, who were intervening on behalf of the companies, on the other. Since almost all miners employed by the state-owned *Gécamines* lost their jobs after its near collapse in the early 2000s, a large part of the former workforce switched to artisanal mining to make a living (Faber et al., 2017). The events around the Tenke Fungurume mine owned by China Molybdenum provide a good example of this type of conflict. In 2010 and 2014, artisanal miners protested against police forces that violently expelled them from the concession of the Tenke Fungurume Mine (TFM). In response, artisanal miners burnt and looted company buildings and machinery (Rubbers, 2019). In June 2019, the Congolese military deployed hundreds of soldiers to yet again prevent artisanal miners from accessing and mining the TFM concession (Ross, 2019).

A second type of conflict played out between the local communities hosting industrial mines and the mining companies. These conflicts focus on the benefits for local communities derived from industrial mining. The most prominent issue in such conflicts is the employment conditions for the local population (CORDAID, 2015). The establishment of industrial mines often raises great expectations among the local population in relation to job creation and the amelioration of infrastructure and general living conditions. Here again, TFM serves as a prominent example. The populations of the neighboring towns Tenke and Fungurume have tripled since the establishment of the mine in 2005. Many immigrants came with the hope of finding a job at the mine. The majority of them, however, ended up working in artisanal mines in the area (Rubbers, 2019). In 2008, several thousand people protested in the city of Fungurume against the lack of employment opportunities for the local population at the Tenke Fungurume mine (Geenen & Cuvelier, 2019). In the following years, the association *Lwanzo Lwa Mikuba*, together with some local chiefs, lobbied for the preferential employment of members of the Sanga ethnic group in industrial mining companies (Gobbers, 2016). *Lwanzo Lwa Mikuba* is a cultural association that represents the interests of the Sanga-speaking community. *Lwanzo Lwa Mikuba* organized marches, wrote letters to, and demanded meetings with representatives of TFM to advocate for preferential employment opportunities for the members of the local population, who they define as all members of the Sanga-speaking tribes. They lament the supposed grip of the Lunda people—a different ethnic group—over the recruitment process (Rubbers, 2019). This shows how closely mining conflicts in the DRC are entangled with long-standing tensions within Congolese society.

Demands for improved employment opportunities for local communities are often put forward together with broader demands related to the benefits host communities can derive from industrial mines. Representatives of local communities frequently complain about the inadequacy of the activities realized through the mining companies' Corporate Social Responsibility (CSR) programs, such as investments in schools, health-care centers, and other local infrastructure. Residents of the municipality of Kangabwa, for example, blocked access to the *Société d'Exploitation de Kipoi* (SEK) mine owned by the Australian company Tiger Resources to demand the signing of a development agreement that obliges the company to invest in important infrastructure for the local community (CORDAID, 2015). In Kolwezi, local residents took the safety manager of a mining company hostage in order to get the company to repair the water supply to the adjacent settlement (Hönke, 2010). Other, less prominent demands put forward in this second type of conflict relate to concerns about environmental pollution, contaminated drinking water, and negative impacts on health in the region (Global Witness, 2004). Local and national NGOs also advocate for better management of mining revenues through the state authorities and fairer redistribution of mining taxes to the affected communities through a reform of mining regulations (CORDAID, 2015). Conflicts around these latter demands so far seem to lead less to physical confrontations. Instead, they tend to be taken up by NGOs who campaign on behalf of the local communities.

Conflicts around cobalt mining: The downstream end in the Global North

At the downstream end of the supply chain, producers of electric vehicles and electronic products have become the focal point of consumer campaigns initiated by various international civil society organizations. NGOs such as Amnesty International and Global Witness denounce an energy transition in the transport sector that is built upon human rights violations and environmental destruction in the Global South. Amnesty International, for example, published the report *Time to Recharge* in 2017, in which the organization evaluates the cobalt sourcing practices of major car manufacturers and electronics producers such as BMW, Tesla, and Apple. Other prominent NGOs such as Global Witness and the Dutch organization Centre for Research on Multinational Corporations (SOMO) likewise published reports and started media campaigns demanding supply chain due diligence from the producers of electric vehicles. In Germany, these demands are supported by the Raw Materials Working Group, a network that comprises several large NGOs such as Brot für die Welt and Germanwatch. The NGOs denounce human right abuses like child labor, as well as social and ecological problems linked especially to artisanally mined cobalt and to a lesser extent to the industrial cobalt mining industry in the DRC (Amnesty International, 2016; Bwenda, 2018; Global Witness, 2004). They demand due diligence from companies involved in the trade, manufacturing, and consumption of cobalt products and campaign for obligatory due diligence through an expansion of the existing European and US regulations on tin, tungsten, tantalum, and gold (3TG), or through new laws and regulations (London Mining Network, 2016) (see Franken et al. within this book). Within these campaigns, the NGOs explicitly frame human rights abuses linked to cobalt mining as a contradiction to the production of green technologies such as electric vehicles, and attack the image of electric cars as sustainable, clean, and green. They demand "ethical batteries to power the clean energy revolution" (Amnesty International, 2017: p. 81).

These due diligence campaigns build upon the very successful "conflict minerals" campaigns of the early 2000s. These campaigns were carried out by various NGOs; the most prominent among them were Global Witness and the Enough Project. The latter was founded by John Prendergast and received support from Hollywood celebrities such as George Clooney (Radley, 2017). The Enough Project used striking slogans such as "No blood in my cell phone" and primarily tackled major producers of electronic consumer goods such as Apple and Samsung. The main narrative of the conflict minerals campaigns organized by international NGOs was that armed groups operating in eastern DRC were violently seizing control of mineral resources that in turn financed their activities. Western consumers were encouraged to exert pressure on the producers of electronic consumer goods to stop purchasing these minerals and so help to end the conflict (Radley, 2017). This narrative is now being reused in the due diligence campaigns, although it has been tailored to cobalt mining. As such, it focuses less on armed conflicts (which are largely absent in Lualaba and Haut-Katanga) and more on ending human rights abuses like child labor. Furthermore, NGOs are not just targeting the producers of electronic goods, but increasingly manufacturers of electric vehicles as well. The narratives have changed to incorporate the importance of minerals for green technologies (e.g., Amnesty International, 2017).

The due diligence campaigns have not yet succeeded in pushing politicians in Western countries to pass legally binding due diligence regulations in the cobalt supply chain. However, in response to the campaigns, the industry has launched the

Responsible Cobalt Initiative, which aims to establish a cobalt supply chain management system in accordance with the Organization for Economic Cooperation and Development (OECD) due diligence guidelines. A total of 31 companies have joined the initiative at the time of writing, including Volvo, BMW, and Daimler. It is too early to assess the results of this initiative.

Conflicts around cobalt mining: Connections and contradictions

Conflicts around cobalt mining take place in the Global South as well as in the Global North. A theoretical perspective combining the concept of global production networks and an action-focused approach to conflicts rooted in the field of political ecology makes it possible to reveal the connections and contradictions of these conflicts. Local communities at the upstream end of the cobalt supply chain are primarily concerned with having access to artisanal mining sites and employment opportunities. The eviction of artisanal miners from industrial concessions frequently sparks violent conflicts in the DRC. Artisanal and small-scale mining constitutes an important livelihood strategy for many people in the DRC. Local communities struggle to maintain access to artisanal mining sites to secure and improve their livelihoods in a period when industrial copper-cobalt mining is expanding.

International NGOs carrying out consumer campaigns for due diligence demand the establishment of mechanisms to hold multinational corporations accountable for human rights violations along their supply chains. They particularly denounce child labor in artisanal mining sites. The idea of holding multinational companies responsible for human rights violations is widely accepted in the DRC as well. However, the strong focus of international NGOs on the working conditions in artisanal mining sites risks partially undermining the demands of local communities for secure access to the artisanal mining sites.

Studies on supply chain initiatives that were established as a result of the conflict minerals campaigns run by international NGOs in the early 2000s in Eastern Congo suggest that supply chain due diligence initiatives have ambivalent effects on the livelihoods of local communities and the income of artisanal miners has decreased in some areas (Radley & Vogel, 2015; Vogel, Musamba, & Radley, 2018; Vogel & Raeymaekers, 2016). There were various reasons for these outcomes. For instance, artisanal mines had to shut down for long periods of time until certification processes were completed, authorities rarely issued clear titles for land and mining, and there was a general lack of technical and financial support for artisanal miners (Geenen, 2012). Furthermore, the top-down formalization process and traceability reforms destroyed many of the existing albeit informal local regulations for artisanal mining sites and frequently created tensions among different newly established mining cooperatives (Vogel et al., 2018).

Conflicts at the upstream end of the cobalt supply chain are also sparked by demands for a fairer distribution of benefits and particularly the incorporation of local communities in industrial mining projects through the creation of employment opportunities. Here, demands made by protest actors in the upstream and downstream ends of the GPN overlap more closely than in the case of artisanal mining. International NGOs, local communities, and national NGOs in the DRC all denounce human rights violations and bad labor conditions in the industrial mines and demand that host communities benefit from the expansion of industrial mining. These issues could act as a starting point to connect the struggles at different points of the supply chain in more strategically effective ways. As the analysis has shown, both artisanal miners and local communities in the DRC and the due diligence campaigns in the Global North exert power in different ways to shape the GPN. Artisanal miners and local communities occasionally manage to interrupt or slow down the cobalt production process through actions such as blocking access roads, destroying mining equipment, or refusing to leave industrial concessions. NGOs in the Global North exert power on the GPN through lobbying for due diligence regulations and publicly shaming well-known consumer brands (such as Volkswagen & Daimler) for using cobalt from the DRC that has been mined using child labor. These big brands have a fairly powerful position in the GPN since they are the major buyers of lithium-ion batteries. In response to the due diligence campaigns, large companies involved in the lithium-ion battery GPN have launched the Responsible Cobalt Initiative, which shows how sensitively companies react to consumer campaigns (see Franken et al. within this book). If the due diligence campaigns prove to be as successful as the conflict minerals campaigns, this might strongly influence the shape of the global production network surrounding lithium-ion batteries.

A more strategic alliance between local communities in the DRC and international NGOs, mediated by Congolese NGOs, for example, could allow for more coordinated protest actions. Such an alliance could also identify logistical choke points and more clearly expose the externalization of the social and environmental costs of the energy transition to the Global South. However, for this to happen, international NGOs must take the demands of artisanal miners and local communities in the DRC seriously, and reduce the unequal power relations between international NGOs and protest actors in the DRC. This might also help to avoid some of the mistakes that were made with regard to the conflict minerals regulation in Eastern Congo.

Conclusion: Conflicts linked to raw materials required for energy transitions

The energy transition is heavily reliant on the secure supply of a variety of metals. Until recycling processes are improved, this will lead to the opening up of new regions for industrial and artisanal mining. Many raw materials that are essential for the energy transition are mined in the Global South. Here, the expansion of the mining frontier often sparks conflicts. In this article, I have argued that the energy transition and more specifically the electrification of the transport sector is related to conflicts around the production of cobalt. The increased demand for cobalt for the production of lithium-ion batteries has contributed to an expansion of industrial and artisanal cobalt mining in the DRC and thus to the physical, economic, and social transformation of the copper-cobalt mining areas. This process is highly contested in the DRC. I identified two key types of conflicts: First, conflicts between artisanal miners and industrial companies over access to artisanal mining sites. Second, conflicts between the local host communities and industrial mining companies over the distribution of the benefits derived from industrial mining, especially with regard to the creation and distribution of employment opportunities. However, as I have shown, the expansion of cobalt mining is not only contested within the DRC itself; conflicts take place at the downstream end of the cobalt supply chain as well, with international NGOs initiating consumer campaigns for the responsible sourcing of minerals and demanding due diligence from car manufacturers such as Volkswagen and Daimler. The relationship between these different struggles is marked by potential points of connection, especially around the distribution of mining profits and stopping human rights violations perpetrated by multinational mining companies. Divisions emerged with regard to artisanal mining. Struggles in the DRC are defensive struggles around access to land for artisanal mining, which constitutes a key source of livelihood, while the due diligence campaigns advocate the regulation and formalization of artisanal mining to end child labor.

In looking at these struggles at different points of the global production network, it becomes clear that the energy transition is linked to a variety of conflicts related to the raw materials required for green technologies. Protest actors contesting the expansion or implementation of industrial mining in the Global South as well as NGOs and social movements in the Global North are important actors in the GPNs of green technologies. Both types of actors can exert pressure on companies and states within the global production network in different ways, and thus help shape the changing relations of production linked to the energy transition.

References

Al Barazi, S., Näher, U., Vetter, S., Schütte, P., Liedtke, M., Baier, M., et al. (2017). *Cobalt from the DR Congo—Potential, risks and significance for the global cobalt market.* Commodity Top News Hannover: Bundesanstalt fuer Geowissenschaften.

Amnesty International. (2016). *"This is what we die for": Human rights abuses in the Democratic Republic of the Congo power the global trade in cobalt.* London.

Amnesty International. (2017). *Time to recharge. Corporate action and inaction to tackle abuses in the cobalt supply chain.* London.

Anlauf, A. (2016). Greening the imperial mode of living? Socio-ecological (in)justice, electromobility, and lithium mining in Argentina. In M. Pichler, C. Staritz, K. Küblböck, C. Plank, W. Raza, & F. Ruiz Peyré (Eds.), *Fairness and justice in natural resource politics.* London: Routledge.

Bebbington, A. & Bury, J. (Eds.), (2013). *Subterranean struggles. New dynamics of mining, oil, and gas in Latin America.* Austin: University of Texas Press.

Bonacker, T., & Imbusch, P. (2006). Zentrale Begriffe der Friedens- und Konfliktforschung: Konflikt, Gewalt, Krieg, Frieden. In P. Imbusch & R. Zoll (Eds.), *Friedens- und Konfliktforschung.* Wiesbaden: VS Verlag für Sozialwissenschaften.

Bridge, G. (2008). Global production networks and the extractive sector: Governing resource-based development. *Journal of Economic Geography*, 8, 389–419.

Bryant, R. L., & Bailey, S. (1997). *Third world political ecology*. London: Routledge.

Bwenda, C. (2018). Go ahead, try to accuse us…. In *Human rights violations by Chinese mining companies in the Democratic Republic of Congo: The case of China Nonferrous Metal Mining Co. in Mabende*. Lubumbashi: SOMO; IPREMI CONGO.

Church, C., & Crawford, A. (2018). *Green conflict minerals: The fuels of conflict in the transition to a low-carbon economy*. Winnipeg: International Institute for Sustainable Development.

Coe, N. M., Dicken, P., & Hess, M. (2008). Global production networks: Realizing the potential. *Journal of Economic Geography*, 8, 271–295.

Coe, N. M., Hess, M., Yeung, H.W.-C., Dicken, P., & Henderson, J. (2004). 'Globalizing' regional development: A global production networks perspective. *Transactions of the Institute of British Geographers*, 29, 468–484.

Conde, M. (2017). Resistance to mining: A review. *Ecological Economics*, 132, 80–90.

Conde, M., & Le Billon, P. (2017). Why do some communities resist mining projects while others do not? *The Extractive Industries and Society*, 681–697.

CORDAID. (2015). *L'exploitation miniere au coeur des zones rurales: Quel developpement pour les communautes locales?* La Haye.

Cumbers, A., Nativel, C., & Routledge, P. (2008). Labour agency and union positionalities in global production networks. *Journal of Economic Geography*, 8, 369–387.

De Koning, R. (2009). *Artisanal mining and postconflict reconstruction in the Democratic Republic of the Congo*. Stockholm: SIPRI.

Dietz, K., & Engels, B. (2014). Immer (mehr) Ärger wegen der Natur?—Für eine gesellschafts- und konflikttheoretische Analyse von Konflikten um Natur. *Österreichische Zeitschrift für Politikwissenschaft*, 43, 73–90.

Dietz, K., & Engels, B. (2017). Contested extractivism: Actors and strategies in conflicts over mining. *Die Erde*, 148, 111–120.

Escobar, A. (2006). Difference and conflict in the struggle over natural ressources: A political ecology framework. *Development*, 49, 6–13.

Faber, B., Krause, B., & Sanchez de la Sierra, R. (2017). *Artisanal mining, livelihoods, and child labor in the cobalt supply chain of the Democratic Republic of Congo*. CEGA White Papers UC Berkeley: Berkeley.

Frankel, T. C. (2016). *The Cobalt Pipeline. Tracing the path from deadly hand-dug mines in Congo to consumers' phones and laptops*. Washington Post, 30.09.2016.

Geenen, S. (2012). A dangerous bet: The challenges of formalizing artisanal mining in the Democratic Republic of Congo. *Resources Policy*, 37(3), 322–330.

Geenen, S., & Cuvelier, J. (2019). Local elites' extraversion and repositioning: Continuities and changes in Congo's mineral production networks. *The Extractive Industries and Society*, 6, 390–398.

Global Witness. (2004). *Rush and ruin. The devastating mineral trade in Southern Katanga, DRC*. Washington.

Gobbers, E. (2016). Ethnic associations in Katanga province, the Democratic Republic of Congo: Multi-tier system, shifting identities and the relativity of autochthony. *The Journal of Modern African Studies*, 54, 211–236.

Gunn, G. (Ed.), (2014). *Critical metals handbook*: John Wiley & Sons.

Haus, R. (2017). *DERA Industrieworkshop Lithium: Batterierohstoffe für Lithiumionenbatterien*. Berlin: DERA.

Hönke, J. (2010). New political topographies. Mining companies and indirect discharge in Southern Katanga (DRC). *Politique Africaine, 120*, 105–127.

IEA. (2018). *Verteilung der energiebedingten CO2-Emissionen weltweit nach Sektor im Jahr 2016. Statista.* Available from: https://de.statista.com/statistik/daten/studie/167957/umfrage/verteilung-der-co-emissionen-weltweit-nach-bereich/. (Accessed August 6, 2019).

Katz-Lavigne, S. (2019). Artisanal copper mining and conflict at the intersection of property rights and corporate strategies in the Democratic Republic of Congo. *The Extractive Industries and Society, 6*, 399–406.

London Mining Network. (2016). *EU agrees mandatory law on conflict minerals, but exemptions cause concern.* Available from: http://londonminingnetwork.org/2016/11/eu-agrees-mandatory-law-on-conflict-minerals-but-exemptions-cause-concern/. (Accessed August 11, 2019).

Marscheider-Weidemann, F., Langkau, S., Hummen, T., Erdmann, L., & Tercero Espinoza, L. (2016). *Rohstoffe für Zukunftstechnologien 2016.* Berlin: Deutsche Rohstoffagentur (DERA).

Martinez-Alier, J. (2009). Social metabolism, ecological distribution conflicts, and languages of valuation. *Capitalism Nature Socialism, 20*, 58–87.

Ngoie, G. T., & Omeje, K. (2008). Rentier politics and low intensity conflicts in the DRC: The case of Kasai and Katanga provinces. In K. Omeje (Ed.), *Extractive economies and conflicts in the global South.* London: Routledge.

Olivetti, E. A., Ceder, G., Gaustad, G. G., & Fu, X. (2017). Lithium-ion battery supply chain considerations: Analysis of potential bottlenecks in critical metals. *Joule, 1*, 229–243.

Prause, L. & Dietz, K., n.d. forthcoming. Die andere Seite des Elektromotors: Konflikte um Rohstoffabbau im Globalen Süden. In: Brunnengräber, A. & Haas, T. (Eds.) Baustelle Elektromobilität. Bielefeld: transcript.

Pye, O. (2017). A plantation precariat: Fragmentation and organizing potential in the palm oil global production network. *Development and Change, 48*, 942–964.

Radley, B. (2017). *Western advocacy groups and (class) conflict in the Congo.* CETRI: Southern Social Movements Newswire. Available from: https://www.cetri.be/Western-Advocacy-Groups-and-Class?lang=fr. (Accessed August 11, 2019).

Radley, B., & Vogel, C. (2015). Fighting windmills in Eastern Congo? The ambiguous impact of the 'conflict minerals' movement. *The Extractive Industries and Society, 2*, 406–410.

Ross, A. (2019). *Congo deploys army to protect China Moly's copper mine from illegal miners.* Reuters. Available from: https://www.reuters.com/article/us-congo-mining-cmoc/congo-deploys-army-to-protect-china-molys-copper-mine-from-illegal-miners-idUSKCN1T-K1HX. (Accessed October 22, 2019).

Rubbers, B. (2019). Mining boom, labour market segmentation and social inequality in the congolese copperbelt. *Development and Change, 44*.

Sanderson, H. (2019). *Congo, child labour and your electric car.* Financial Times. Available from: https://www.ft.com/content/c6909812-9ce4-11e9-9c06-a4640c9feebb. (Accessed October 15, 2019).

Scholten, D. (Ed.), (2018). *The geopolitics of renewables*: Springer Nature.

Sovacool, B. K. (2019). The precarious political economy of cobalt: Balancing prosperity, poverty, and brutality in artisanal and industrial mining in the Democratic Republic of the Congo. *The Extractive Industries and Society, 6*(3), 915–939.

Sun, X., Hao, H., Liu, Z., Zhao, F., & Song, J. (2019). Tracing global cobalt flow: 1995–2015. *Resources, Conservation and Recycling, 149*, 45–55.

UNEP. (2009). *From conflict to peacebuilding. The role of natural resources and the environment.* Nairobi.

US Geological Survey. (2009). *Mineral commodity summaries.* U.S. Department of the Interior. Available from: https://s3-us-west-2.amazonaws.com/prd-wret/assets/palladium/production/mineral-pubs/mcs/mcs2011.pdf. (Accessed July 8, 2019).

US Geological Survey. (2019a). *Mineral Commodity Summaries.* U.S. Department of the Interior. Available from: https://s3-us-west-2.amazonaws.com/prd-wret/assets/palladium/production/mineral-pubs/mcs/mcs2011.pdf. (Accessed July 8, 2019).

US Geological Survey. (2019b). *Cobalt reserves worldwide as of 2018, by country (in metric tons). Statista.* Available from: https://www.statista.com/statistics/264930/global-cobalt-reserves/. (Accessed July 8, 2019).

Vogel, C., Musamba, J., & Radley, B. (2018). A miner's canary in eastern Congo: Formalisation of artisanal 3T mining and precarious livelihoods in South Kivu. *The Extractive Industries and Society, 5,* 73–80.

Vogel, C., & Raeymaekers, T. (2016). Terr(it)or(ies) of peace? The congolese mining frontier and the fight against "conflict minerals". *Antipode, 48,* 1102–1121.

Wellmer, F. W., Buchholz, P., Gutzmer, J., Hagelueken, C., Herzog, P., Littke, R., & Thauer, R. K. (Eds.), (2019). *Raw materials for future energy supply*: Springer Nature.

Ziemann, S., Grunwald, A., Schebek, L., Müller, D. B., & Weil, M. (2013). The future of mobility and its critical raw materials. *Revue de Métallurgie, 110,* 47–54.

Voluntary sustainability initiatives: An approach to make mining more responsible?

11

Gudrun Franken[a], Laura Turley[b], and Karoline Kickler[a]

[a]Bundesanstalt für Geowissenschaften und Rohstoffe, Hannover, Germany
[b]International Institute for Sustainable Development & University of Geneva, Geneva, Switzerland

Introduction

The transition to renewable energies is positioned to lead to an increase in global demand for metals and minerals. Increased demand, compared to current production levels, is anticipated for cobalt, lithium, and rare earth element extraction (Marscheider-Weidemann et al., 2016), which are needed for battery technologies for electric vehicles as well as magnets, for example, wind power generation (see Wang et al. within this book). Several of these minerals have been listed as critical elements by the European Union (EU), meaning they are vital for the economic development in the EU and at the same time their supply might be critical due to market concentration of producing countries and/or governance risks in these countries. In addition, the construction of infrastructure for renewable energy technologies such as windmills, transmission lines solar power plants also requires major metals such as copper, iron/steel, and aluminum as well as industrial minerals such as silica. For example, building a 3-MW windmill turbine requires 4.7 tons of copper, 335 tons of steel, 1200 tons of concrete and 3 tons of aluminum (World Bank, 2019) (see Zepf within this book). As such, demand for these base metals for construction is also expected to soar (Vidal, Goffé, & Arndt, 2013). Moreover, extracting these metals and minerals is also an energy-intensive activity. The Global Resources Outlook 2019 (Oberle et al., 2019) states that around 20% of global climate change impacts is currently attributable to the extraction of metals and industrial minerals. So while energy transitions toward more renewable energy technologies are driven by concern about carbon emissions and the climate crisis, mineral production itself has an environmental footprint, especially related to energy consumption but also to water demand and harmful emissions (McManus, 2012), for example, in the case of lithium mining in Latin America (Munk et al., 2016).

With respect to social issues, the positive impacts of mining such as income and employment generation, and local economic development, can be juxtaposed with the negative impacts that have also come to define the sector, such as human rights violations, occupational health and safety risks or resettlement. For commodities produced partly by artisanal and small-scale mining (ASM), child labor and poor working conditions have been documented, for example, in the cases of cobalt (Amnesty International Ltd, 2016).

In the global sustainable development goals (SDGs), negative and positive feedback loops are an important theme. In this sense, improvements toward combatting climate change (SDG 13) should not lead to by an increase in unsustainable production patterns of the technologies needed (SDG 12). Therefore, the need to ensure mining is done in a responsible and resource-efficient manner and it is imperative if we do not want to replace one evil with another.

Life cycle assessments (LCAs) are one way to measure and compare dimensions of sustainability in mining and can factor in differences in technologies used, and thus effects of substitution and technology transition. However, LCA is oriented mostly toward evaluating material inputs, emissions, and impacts along supply chains for specific commodities. The method is not suitable, or does not intend to assure downstream users that commodities are produced in a sustainable way more generally, by reporting on, for example, human rights violations or working conditions in supply chains.

One established way to take action and communicate more sustainable practices in extraction, production, and processing phases is through the use of voluntary sustainability initiatives (VSIs).[a] As the name implies, these are nongovernmental voluntary initiatives that companies and individual mine sites may join. The VSI approach is already well-established in many agricultural commodities, such as in coffee, tea, cacao, cotton, and sugar. In these agricultural sectors, and others, we find certification schemes including product labeling (Rainforest Alliance, Fairtrade, UTZ certification, etc.), and in some cases, the rate of annual growth of certified or "standard-compliant" production is outstripping the growth of conventional production, for example, in coffee (Voora, Bermudez, & Larrea, 2019).

In the mining and minerals sector, a broad range of such initiatives has been developed, particularly since the 2000s, to provide evidence for buyers of raw materials, financial institutions, or the public that certain sustainability criteria are being met. Some of the earliest voluntary normative frameworks in the mining sector were actually government-led, multi-stakeholder initiatives, such as the Voluntary Principles for Security and Human Rights in 2000, the Extractive Industries Transparency Initiative (EITI) in 2002, and the Kimberly Process Certification Scheme in 2003. Both the Kimberly Process (for diamonds) and the International Cyanide Management Code (for gold) were founded in 2003 in response to internationally reported human rights violations in mining and mineral trade.

---

[a] VSS, or voluntary sustainability standards, is another term used frequently to describe the same phenomenon. We choose the term initiatives to be more inclusive to various organizational designs, some of which are more of a community of interest, than strictly issuing standards.

Around the same time, concerns about the risks of mining operations to local communities, as well as the degradation of ecosystems triggered the mining industry itself to develop sustainability initiatives and standard systems. As a result of the global multi-stakeholder project mining, minerals, and sustainable development, the International Council of Mining and Metals (ICMM) was established with their sustainable development principles published in 2003. The Mining Association of Canada's (MAC) Towards Sustainable Mining (TSM) commitments followed in 2004.

Going beyond the promotion of principles to actual certification, the Oro Verde Initiative in Colombia was the first regional standard for the certification of sustainable, artisanal gold (Potts, Wenban-Smith, Turley, & Lynch, 2018). By 2004 Oro Verde had inspired the creation of the Alliance for Responsible Mining (ARM) with a mandate for the creation of a global standard for the artisanal sector, which resulted in the Fairmined standard in 2010.

The VSI "landscape" in mining and minerals has continued to evolve. Concern about minerals from conflict-affected and high-risk areas, so-called conflict minerals has been a recent, notable driver. The conflict in the Eastern Democratic Republic of the Congo (DRC), where mining and mineral trade of tin, tantalum, tungsten, and gold (the "3TG") have reportedly contributed to the financing of armed groups, has fueled discussions on responsibility in mineral supply chains. As a result, legislation on reporting with regard to conflict minerals became an obligation for US stock exchange listed companies (Dodd Frank Act § 1502) in 2011. In the same year, the OECD launched its guidance related to due diligence in mineral supply chains from conflict-affected and high-risk areas (OECD, 2016). The guidance formulates due diligence requirements for the whole supply chain, addressing upstream (from extraction to the smelter) and downstream (from the stage following the smelter to the final product) actors, respectively. It is an internationally recognized benchmark on good practice for companies. Responding to those new developments, several VSIs have emerged to demonstrate due diligence and compliance with conflict-free 3TG requirements.

This chapter seeks to shed light on the landscape of VSIs in the mining and minerals sector and to discuss what they currently contribute to sustainable development, and what their limitations with regard to responsible mining might be. We draw heavily on our own research from two in-depth reports on the topic, namely a comparative overview of sustainability schemes for minerals (Kickler & Franken, 2017) that was part of the project NamiRo, and the Standards and Extractive Economy report, published by the Sustainability Standards Initiative (Potts et al., 2018). The remainder of the chapter is structured as follows: we first provide a nonexhaustive overview of the many VSIs in mining and minerals that exist today and propose some ways to understand or categorize them. Next, we describe how these VSIs vary by their content, design, assurance systems, and other features, and what these differences imply for sustainable development. Finally, we provide insight into the achievements and limitations of this approach to date. The conclusion suggests some considerations specifically related to the energy transition.

The landscape of voluntary sustainability initiatives
Overview of initiatives

Numerous sustainability initiatives and standards in the mining sector exist, and this section seeks to provide some ways to understand and categorize them. The field is certainly characterized by continuous change. VSIs that have existed for some time are continually developing their standard requirements and processes. Moreover, many VSIs have emerged just in the last decade, in some cases in response to the demand for new metals and minerals that have become important in global supply chains. We expect this evolution to continue, as many new initiatives are still in their infancy, for example, with regard to responsible copper or responsible cobalt. Only recently, a report by Amnesty International on child labor in cobalt mining in the DRC (Amnesty International Ltd, 2016) has again been a driver for the formation of four new initiatives for responsible sourcing of cobalt. For copper, the International Copper Association launched The Copper Mark in April 2019—an assurance process for sustainably produced copper.

If we take into account all initiatives from recent decades aiming at fostering sustainability, Potts et al. (2018) identified as many as 158 different standards or initiatives. However, narrowing that down to initiatives specific to the mining and mineral sector that have developed standard frameworks, with related governance systems, leaves around several tens of active initiatives. Potts et al. (2018) studied and compared 15 VSIs in detail, and Kickler and Franken (2017) came up with a similar number of 19 relevant sustainability schemes addressing mining and metals explicitly. Table 11.1 gives an overview of some of the key sustainability initiatives in the mining sector and provides the full names for the acronyms used extensively throughout the chapter.

Table 11.1 Sustainability schemes in mining, organized by the mineral commodity they address.

Mineral commodities	Sustainability schemes	Responsible organization	Abbreviation
All mineral resources	Environmental and Social Performance Standards	International Finance Corporation (IFC)	IFC
	Standard for Responsible Mining	Initiative for Responsible Mining Assurance (IRMA)	IRMA
	Toward Sustainable Mining (TSM)	Mining Association of Canada (MAC)	MAC
	Sustainable Development Framework (SDF)	International Council on Mining and Metals (ICMM)	ICMM
	GRI Reporting Principles and Standards Disclosure and Sector Supplement	Global Reporting Initiative (GRI)	GRI

Table 11.1 Sustainability schemes in mining, organized by the mineral commodity they address—cont'd

Mineral commodities	Sustainability schemes	Responsible organization	Abbreviation
Gold	International Cyanide Management Code (Cyanide Code) For the Manufacture, Transport, and Use of Cyanide In the Production of Gold	International Cyanide Management Institute (ICMI)	Cyanide Code
	Conflict Free Gold Standard (WGC)	World Gold Council (WGC)	WGC
	LBMA Responsible Gold Guidance	The London Bullion Market Association (LBMA)	LBMA
Gold and associated silver and platinum	Fairmined Standard for Gold from Artisanal and Small-scale Mining, including Associated Precious Metals	Alliance for Responsible Mining (ARM)	Fairmined
	Fairtrade Standard for Gold and Associated Precious Metals for Artisanal and Small-Scale Mining	Fairtrade Labeling Organizations International e.V. (FLO)	Fairtrade
Gold, tin, tantalum and tungsten	Responsible Minerals Assurance Program	Responsible Minerals Initiative	RMAP
	Certified Trading Chains (CTC); adapted by the DR Congo	The Ministry of Mines of the Democratic Republic of Congo (DRC)	CTC
Tin, tantalum, tungsten	Regional Certification Mechanism (RCM)	Regional Initiative against Illegal Exploitation of Natural Resources (RINR)	RCM
	ITRI Tin Supply Chain Initiative (iTSCi) membership program agreement summary (only for 3T)	International Tin Research Institute (ITRI)	iTSCi
Diamonds, gold and platinum[a]	RJC Code of Practices and RJC Chain-of-Custody Standard	Responsible Jewelry Council (RJC)	RJC
Aluminum	ASI Performance Standard and ASI Chain-of-Custody Standard	Aluminum Stewardship Initiative (ASI)	ASI

Continued

Table 11.1 Sustainability schemes in mining, organized by the mineral commodity they address—cont'd

Mineral commodities	Sustainability schemes	Responsible organization	Abbreviation
Natural stone	Fair Stone—International Standard for the Natural Stone Industry	Fair Stone e.V.	Fair Stone
	XertifiX Criteria	XertifiX e.V.	XertifiX
	Responsible Aggregate Standard	Cornerstone Standards Council	CSC
	Natural Stone Sustainability Standard	Natural Stone Council	NSC
Coal	Bettercoal Code	Bettercoal Initiative	Bettercoal

[a] *Platinum group metals.*
Credit: Kickler, K., & Franken, G. (2017). Sustainability schemes for mineral resources: A comparative overview, Hannover: Bundesanstalt für Geowissenschaften und Rohstoffe, p. 167 and Potts, J., Wenban-Smith, M., Turley, L., & Lynch, M. (2018). Standards and the extractive economy, International Institute for Sustainable Development. ISBN: 978-1-894784-79-5. https://www.iisd.org/ sites/default/files/publications/igf-ssi-review-extractive-economy.pdf (Accessed 15 July 2019).

Most initiatives that address exploration and mining have general, not commodity-specific, requirements and refer to industrial mining. The most prominent being ICMM, GRI, IFC, MAC, and IRMA. These initiatives cover a broad range of sustainability issues whereas the Cyanide Code and the WGC Standards address only selected issues (cyanide management or conflict-related issues, respectively). Others, however, can only be applied to the artisanal mining sector such as Fairmined or Fairtrade, or refer only to the smelter/refinery level such as LBMA and RMAP.

Due to the inherent diversity of the mining and minerals sector, VSIs can be grouped or classified along many different potential axes. One way to assess and categorize VSIs is by defining at what stage of the supply chain they are active (see Fig. 11.1). For some VSIs, standards apply to only certain parts of the supply chain, but compliance is communicated along the whole supply chain, for example, from mining to smelter or from mining down to the end consumer. Examples are the Fairmined or Fairtrade Standard for gold, silver, and platinum, but also natural stone schemes, such as XertifiX or Fair Stone. Examples of upstream schemes that cover mine to export are those developed in the African Great Lakes Region, against the background conflict-financing in the DRC such as CTC, iTSCi, and RCM.

Initiatives can also be categorized based on characteristics of the mining production site they apply to such as industrial large-scale, artisanal, and small-scale (ASM), and specific aggregate and quarrying mines. Beyond this, some standards are applicable with limited scope based on geography, the commodity-type or the "issue" areas of concern (e.g., child labor, reduction of greenhouse gas emissions, tailings management, etc.).

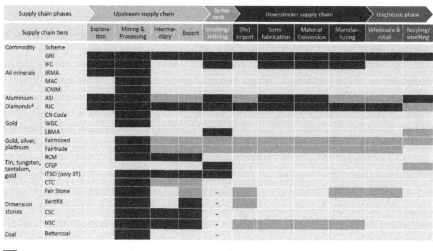

FIG. 11.1

Schemes and their requirements along the mineral (model) supply chain. See Table 11.1 for nonabbreviated references.

Modified from Kickler, K., & Franken, G. (2017). Sustainability schemes for mineral resources: A comparative overview, Hannover: Bundesanstalt für Geowissenschaften und Rohstoffe, p. 167.

VSIs may also be distinguished by the level of obligation they impose on companies or mine sites (Potts et al., 2018). Some initiatives are designed as standards that are more or less obligatory. Some examples of this "full compliance" approach at the time of writing are RJC, ASI, IFC, CSC, Fairtrade, and Fairmined. In these cases, standard-complaint or certificate-holding entities are expected to meet all specified requirements. Other initiatives may be designed with staggered levels of achievement with bronze, silver, and gold levels, for example, IRMA, or with some compulsory and some optional criteria (e.g., Xertifix), or with high levels of discretion regarding what constitutes compliance (e.g., ICMM). Finally, others may not be in the business of certifying practices at all but will be more of a reporting tool (e.g. TSM). As described in Potts et al. (2018), these differences are not inherently "good" or "bad" but are just different.[b] They may reflect, at least in part, different theories of change at work. For example, VSIs that opt for a more flexible approach to compliance may consider that it is important not to exclude "poor performers" a priori, but to find a way to bring them into a system of sustainability and encourage continuous improvement through levels or requirements that become more stringent over time.

---

[b] The characterizations of VSIs given are accurate at the time of writing but are likely to change and evolve over time.

Content of sustainability requirements

Sustainability requirements that are addressed by VSIs differ depending on the objective of the scheme, the mineral it might address as well as on the drivers that led to the development of the scheme. For example, small-scale mining is known for its precarious working conditions. Thus, not surprisingly, human rights and social issues are a major focus of schemes such as Fairmined or Fairtrade. Similarly for 3TG mining, standards typically focus on addressing the risks related to human rights and conflict financing in conflict-affected and high-risk areas. Environmental issues are barely addressed (yet) in these schemes. However, we see a continuous evolution of VSIs. For example, whereas the RMAP only provides certification against issues addressing conflict-financing and severe human rights violation (in-line with the risks laid down in the OECD due diligence guidance for conflict-affected and high-risk areas), the RMI, the organization responsible for the RMAP, has developed a self-assessment tool for companies. It goes beyond conflict and addresses additional areas such as labor and working conditions, environment, and community issues. An overview of the different issues that sustainability schemes in mining address can be found in Fig. 11.2, from Kickler and Franken (2017). To compare the requirements of various schemes, a catalog of 5 categories with 14 issues and 86 subissues was developed that covers all aspects addressed in the respective 19 schemes that were analyzed. The structure of the catalog is oriented toward the ISO 26000 standard. It can serve as a reference catalog for existing and new schemes and has been used already for the concept of a new standard, the CRAFT standard (Alliance for Responsible Mining, 2018).

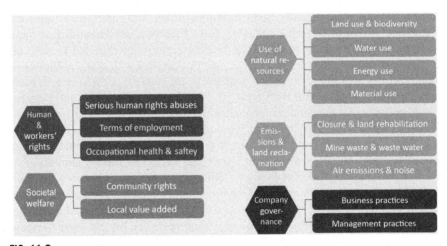

FIG. 11.2

Categories and issues addressed by sustainability schemes.

Modified from Kickler, K., & Franken, G. (2017). Sustainability schemes for mineral resources: A comparative overview, Hannover: Bundesanstalt für Geowissenschaften und Rohstoffe, p. 167.

The analysis of sustainability initiatives in mining against these 86 subissues (grouped in 14 issues; Fig. 11.2) shows that employment conditions, community rights, and occupational health and safety are the most prominent issues addressed by the VSIs overall, followed by human rights, land use and biodiversity, waste, and business management practices. This reveals that standards predominantly address high-risk areas that also pose risks to business reputation and the social license to operate, such as human rights abuses or biodiversity issues (Fig. 11.3).

The reduction of greenhouse gas emissions is likely to become a more prevalent issue category, in response to the climate crisis. Moreover, it will increasingly be important for mines and companies to go beyond the reporting of emissions, to actually improving practices. On this point, Maennling and Toledo (2018) describe efforts made by some voluntary initiatives. Most notably, IRMA has an entire chapter on greenhouse gas emissions, and certification is dependent on having mine-specific targets in place that will be measured against and publicly reported. Others put more emphasis on reporting. ASI's standard number 5 promotes the reduction of CO_2 emissions and the purchase of renewable energy for the smelter process. TSM has developed a protocol that guides member companies on evaluating emissions that explicitly promotes renewable energy investment. Maennling and Toledo (2018) suggest that ICMM strengthen their Principle 6 on environmental performance, to explicitly highlight renewable power integration for reducing emissions on-site.

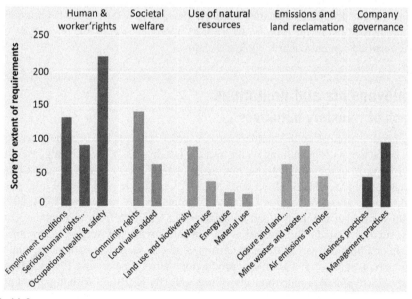

FIG. 11.3

Cumulated scores of issues addressed by sustainability schemes at the mining level. Score relates to the extent of requirements per 86 subissues.

Modified from Kickler, K., & Franken, G. (2017). Sustainability schemes for mineral resources: A comparative overview, Hannover: Bundesanstalt für Geowissenschaften und Rohstoffe, p. 167.

Assurance systems

There are different approaches to how VSIs prove that the parameters defined in their standard are actually being adhered to in practice on mine sites or along the supply chain. This is known as the assurance system that is in place, and it applies specifically to VSIs that issue some form of certification or label for commodities or end products. Generally, initiatives that require independent, third-party verification are considered the more reliable and more rigorous approach to sustainable development vs. ones based only on self-reporting.

Most initiatives discussed in this chapter do use some form of third-party assurance, and there is a trend for new initiatives to do the same. Potts et al. (2018) found that the most common form of assurance system covering large-scale, industrial mining is that of independent third-party auditing, followed by the initiative *itself* determining compliance and issuing a certificate. Unlike in other commodity sectors, the model of a third party that both audits and issues a certificate third party seems to be uncommon in this sector (NSC is an exception).

Closely related to the assurance system is the type of traceability program in place. Traceability refers to the ability to ensure that downstream claims being made about the origins of a product are accurate. Downstream users and brands generally face the most pressure to have traceability programs in place, to show their end consumers that, for example, they are not sourcing from conflict zones, or from certain mines where accidents have occurred. Both the RJC and the ASI have full traceability systems from the mine site through to the end user. ISEAL (2016) provides a comprehensive list of the types of traceability approaches available (e.g., mass balance, certified content control, book and claim).

Achievements and limitations
Reach of voluntary initiatives

In general, participation in voluntary sustainability schemes has become more common in the international mining sector over the last decade. ICMM, the biggest initiative with most of the multinational mining companies, covered a total value of mine production of around 213 billion USD in 2017, which is around 30% of total global production of about 700 billion USD (not counting energy resources) (BGR, 2019).

Most member companies and associations of VSIs are from anglophone countries, such as Canada and Australia. As a result, hardly any mining operations in Russia or China are part of these initiatives, although these two nations account for an important share of global mining production, with China around 19% and Russia around 6% of global production value of minerals (BGR, 2017). With important mining regions such as China or Russia underrepresented in these initiatives, much less is known publicly about sustainability issues and standards of mining in these regions. For example, a nonexhaustive analysis of public information for the 2013–2018 period shows that companies in China reported only approximately half as many social and environmental incidents (spills, strikes, etc.) that were reported in Chile, despite

China's mining sector being about four times bigger than that of Chile (BGR, 2017; Kühnel et al., 2019).

Where China dominates the market of certain commodities such as rare earth elements, it is a challenge from the downstream perspective to establish international sustainability requirements in the political context. Rather, international requirements seem to be adapted to national systems, as the example of the OECD due diligence guidance shows. Though China is not a member of the OECD, it has, with support of the OECD, included the guidance into a Chinese due diligence standard in the mineral and metals sector (CCCMC, 2016). The plan to shut down more than 1000 mines in China due to social and environmental concerns by the end of 2019 shows that there are serious concerns about the performance of some companies. It also shows the preference for a national or state-centric approach to dealing with these matters. However, some Chinese companies are active in voluntary initiatives, specifically where outbound investment or supply chains are concerned such as in the RMAP or in the responsible cobalt initiative.

However, and importantly, a company's membership in a VSI does not necessarily mean all of the company sites are meeting the same standards. Only recently, companies are engaging in site-specific reporting through initiatives such as TSM and IRMA. In the future, the site-level reporting and certification approach might enable more robust data on real market coverage of sustainability standards.

In the artisanal and small-scale (ASM) mining sector, however, we observe very limited market penetration: the Fairmined initiative has sold approximately 1 ton of gold since 2014 (Alliance for Responsible Mining, 2019); Fairtrade has sold less at only several hundred kilograms of gold. Meanwhile, the global mine production of gold was around 3300 tons in 2018 (World Gold Council, 2019). Global artisanal gold production is estimated at around 400 tons (Seccatore, Veiga, Origliasso, Marin, & De Tomi, 2014), which illustrates that the share of certified gold in that sector is negligible. Especially for ASM producers, a significant barrier is the lack of capacity (management and financial) to improve performance. Moreover, incentives are missing due to lack of demand of downstream companies in the gold sector. Major barriers for the uptake of these standards are the relatively high costs for small amounts of material, a less steady and predictable supply compared to industrial mining, and prefinancing challenges due to missing proven geological reserves at ASM sites (Gronwald, 2019).

Harmonization of standard requirements

Another result of the growing number of initiatives is the increasing referencing, harmonization and cross-acknowledgement of standards, without the actual merging of initiatives. References to other international standards are common in the VSIs reviewed—to give one example, ICMM is benchmarked against ILO Conventions, OECD documents and Global Reporting Initiative (GRI) standards. More recently, the Copper Mark initiative references the RMI's Risk Readiness Assessment—an existing standard—but communicates its own governance approach.

The uptake of the OECD due diligence guidance for mineral supply chains from conflict-affected and high-risk areas resulted in voluntary initiatives that build upon another in the supply chain and also have a high level of acknowledgment between initiatives: Fairmined, Fairtrade, RMAP, RJC, LBMA, and iTSCi reference and/or acknowledge each other with regard to the OECD requirements.

Going beyond cross-referencing, a recent collaboration involving Responsible Steel, the RJC, IRMA, and MAC aims to create an *interoperability platform* for VSI in the minerals, mining and metals sector, with an initial focus on coordinated engagement to stakeholders (business and civil society, up and down supply chains), and initial pilot work in South Africa.[c]

While it is still early days, mutual acknowledgment across VSIs is increasing, reflecting perhaps the burden of multiple audits and standards. However, while we see that companies are mainstreaming a common understanding of responsible mining and supply chains in general management frameworks, in practice companies still have many different frameworks to follow. Interviews with members of the TSM initiative (personal communication) revealed, that members of the TSM initiatives report against nine sustainability frameworks on average, some even against up to 15.

From our perspective, interoperability and cross-referencing is a positive trend, especially in as far as it can streamline procedures and costs for target beneficiaries at mines sites, in mining communities and within mining companies. A future development toward a single internationally recognized reference standard framework would be desirable, but at present, it remains unlikely given the huge spectrum of global supply chains and sectors as well as the differing expectations (high-end standard vs. minimum standard).

Transparency of mining and mineral supply chains

As discussed, reporting on sustainability performance is a key characteristic of many VSIs. In some cases, companies already have a legal obligation to report, through the EU regulation on due diligence obligations for importers of tin, tantalum, and tungsten, their ores, and gold originating from conflict-affected and high-risk areas (EU, 2017), or the US Dodd Frank Act § 1502 on conflict minerals. In these cases, VSIs can be useful for demonstrating compliance with regulations. Indeed RMI was set up to communicate conflict-free mineral production and trade along the supply chain from smelters (and, implicitly, the associated mines) to the Original End Manufacturer (OEM). The initiative has generated a global list of certified smelters—the bottleneck between the numerous upstream (mine to smelter) and downstream (manufacturing) companies—and today it is crucial for demonstrating compliance with the relevant regulations. Currently, it is also extending its work toward cobalt producers.

---

[c] Funded by ISEAL's Innovation Fund.

Through these initiatives, we note a modest increase in the transparency of mineral supply chains that have traditionally been quite opaque. For example, the IPIS mapping project monitors and assesses several thousand mine sites in the Eastern DRC. It is supported by the UN, BGR, and local governments, and has resulted in an interactive, continually updated map of actors and mine sites, showing the involvement of armed groups (IPIS, 2019). Another example is an industry initiative on the bagging and tagging of tin, tantalum, and tungsten ore exports, and related risk monitoring on the ground—providing detailed information mineral trade. The analysis of the tantalum and tin concentrate exports from the DRC shows that following the disengagement of international buyers from the Eastern DRC around 2010–2012 (after the enactment of the Dodd Frank Act) trade picked up in the following years when progressively expanding supply chain initiatives established traceability and risk monitoring (Schütte, 2019), thus enabling responsible buyers to engage in the region again. However, these developments were also influenced by other, fundamental market forces, and should not be attributed exclusively to sustainability initiatives.

To communicate responsible performance, the major target groups of most initiatives are business partners and investors who seek to ensure responsible practice and minimize risks in their supply chains or investments. A second major target group is the local public. Maintaining the social license to operate and social acceptance is one of the major business risks in mining (Ernst & Young, 2018). For example, TSM was taken-up in Finland following the Talvivaraa dam failure in 2012, with the major goal to restore public confidence in the local mining industry.

Labels to inform end consumers about responsible production are still rare for minerals and metals. Products, where labeling is applied currently in the sector, are mostly those with only one or few commodities to be certified (e.g., gold, diamonds in the jewelry sector or natural stone). During the last years more complex products with high consumer identification such as smartphones were labeled with respect to some metals and this trend might also be expected in other sectors, for example, automotive given the increasing engagement of this industry in initiatives (see Sonnberger within this book).

Although while labeling is not widely demanded (yet) at the product level, branding is highly important at the company level. This refers, in particular, to mitigating reputational damage.

Looking ahead, data management and data exchange will continue to be key issues for sustainability initiatives. Large mining companies, as well as downstream industries, are establishing or advancing new forms of data exchange procedures including through decentralized database management such as the blockchain approach. The RMI initiative has released guidance on applying blockchain in mineral supply chains (RMI, 2018). The benefits are that information can be communicated to participants without being manipulated, and while respecting business confidentiality. However, even the most sophisticated data exchange procedures still face the major challenge of the accuracy of information initially entered into the system. This can only be achieved through effective on-the-ground monitoring and capacity-building.

Sustainability reporting and performance

Despite increased reporting, there is still the question of how reporting reflects performance. The RMI rates 30 major international mining companies on sustainability, based on publicly available information. The ratings show, for example, that in terms of environmental responsibility, 14 out of the 16 ICMM member companies are ranked in the upper half of the rating. This could reflect the comprehensive reporting standards of the initiative. On the other hand, a company like Vedanta Resources that has had several environmental incidents reported (Divekar, 2018), also ranks fairly high at rank 16 out of 30. Also, the fairly high environmental responsibility ranking of the company Vale (rank 12 out of 30) contrasts with disastrous tailings dam failures. Hence, there is an increased pressure toward independent monitoring of company performance, and VSIs are increasingly introducing independent monitoring and auditing of their members. One recent effort, mainly driven by concerned investors, is by the Principles for Responsible Investors (PRI), the United Nations Environment Programme (UNEP), and ICMM to introduce a global tailings dam monitoring system that is independent of mining companies (PRI, 2019).

Limited capacity to deal with deviant and illegal behavior

Voluntary initiatives for mineral supply chains have generally been setup to address troublesome "hot spots" in supply chains, such as human rights violations or child labor. Whereas some of the initiatives have reached wide market penetration on a global scale, they still depend on the willing participation of operators. If we take certification as an indicator of willing participation, RMI's certification of tin, tantalum, and tungsten smelters has obtained an impressive 90% global penetration (RMI, 2019). However, for gold refineries, their certification program seems to stagnate around 70% (RMI, 2019). This reveals that there is still business for noncertified smelters, and that many buyers do not question the origin of materials.

Moreover, the vast majority of artisanal gold from Eastern DRC continues to be smuggled out of the country, mainly to Dubai (UN, 2019). A current estimate calculated this illegal trade to more than 99% (Neumann et al., 2019). This underlines that illegal practice cannot be addressed by voluntary schemes as long as business opportunities for illegal options remain and governance in the region is weak.

Challenges in the ASM sector, in particular, demonstrate that much more is needed than just voluntary initiatives to promote responsible mining practice. Basic ASM sector formalization and legalization cannot be tackled through a supply chain approach, but instead calls for the important role of the state regulation, monitoring, and enforcement. In cases where the state does not provide an adequately enforced legal framework, it is difficult to establish internationally legitimate ASM supply chains. Furthermore, responsible ASM mining practice needs to tackle a broad range of issues from environmental (e.g., using mercury for gold processing) to social (e.g., child labor), and setting the right priorities and sequencing them may be a challenge for external stakeholders. Positive change will not happen overnight, and long-term strategies are required for progressive improvement.

The cost of certification and voluntary initiatives

Costs for assuring upstream due diligence and certification from mine to export remain mostly with the respective countries and producers in the region. For example, external due diligence costs in Rwandan 3TG supply chains are estimated at 3%–5% of the export value (Cook & Mitchell, 2014). While traceability schemes can provide reliable market access for validated sites, the burden of additional costs for producers in the region is still highly debated.

Once again, the challenges for ASM are particularly acute. ASM can be vital for local income; for instance, in the Eastern DRC, an estimated 200,000–300,000 miners operate in the local 3TG sector. Where VSIs are active in ASM, there is still limited empirical evidence that they have improved working conditions, or enhanced financial and technical capacities of miners, beyond a few pilot projects. The challenge is magnified by the fact that in ASM contexts, land titles to mineral reserves are often not recognized officially, or are not legitimate. As such, the additional, upfront costs associated with responsible sourcing for these outfits may not be feasible without the support of external funds. A recent study of the Eastern DRC even claims that "…increasing regulation of the artisanal mining sector and responsible sourcing efforts, have rather had a negative overall effect on the socioeconomic position of artisanal miners" (Matthysen, Spittaels, & Schouten, 2019). Certainly, the limitations of VSIs to address the marginalized ASM sector and contribute to its development is a major criticism being raised (Hilson, Hilson, & Mc Quilken, 2016).

Integrating additional costs is and will remain a challenge for industrial mining too. In general, future research is needed to segregate and define precisely the sources of the additional costs (e.g., related to on-site improvements, salaries, safety equipment, compliance costs such as auditing, downstream costs like marketing and developing partnerships, research, etc.). Only then can we address them effectively and direct investment efficiently.

Conclusion

Voluntary sustainability initiatives have become players in the global governance of the mineral and metal extractive sector. Looking forward, it seems likely that most industries will use such systems in the future to demonstrate compliance and to be in line with international developments on sustainability standards. The social license to operate will continue to be a significant driver for companies to adopt VSIs. Also, legislative efforts might increasingly refer to general principles and standards addressed in these schemes or even explicitly acknowledge schemes, for example, through procurement regulations. It waits to be seen, however, how initiatives can expand regionally, like into China where a strong regulatory framework controls company development and performance.

The basis of supply chain initiatives is the communication of information along supply chains. New technologies will develop further and probably increase access also to sustainability information beyond company reporting. However, it is not clear how the huge gap between the professional management of international companies and the artisanal mining sector and other small actors of the supply chain can be bridged. Or, if small and/or poor performers will be marginalized to (remaining) markets that do not question production practices.

It seems that expectations on the potential impact of VSIs to spur positive development in difficult environments such as high-risk and conflict-affected areas have been too high. They can support development to some respect but have little effect on the political and economic setting of the sector. Therefore, more emphasis needs to be put on linking voluntary initiatives to government authorities, like through national platforms of the extractive sector such as the Extractive Industries Transparency Initiative (EITI), to communicate performance and shortcomings in the sector. Also, initiatives might help to develop local capacities, for example, as has been done for mine site inspection by the CTC approach. If ambitious standard requirements only lead to market differentiation between first in class performers and all the others, the initiatives are missing out their sustainable development potential, notably related to the SDGs.

With regard to renewable energy technologies, lithium, nickel, and rare earth minerals need to be addressed more by these initiatives, while not forgetting the importance of basc metals to new technologies and infrastructure either. Argentina, a lithium producer, adopted the TSM framework in 2017, together with Botswana, the Philippines, and recently Spain. The Economic Commission for Latin America and the Carribean (ECLAC) under its sustainable development agenda also is reflecting on potentially useful sustainability indicators in mining for the region. If voluntary initiatives could contribute to a common international understanding and implementation of responsible mining practice, that would be a good way forward.

Acknowledgments

The authors would like to thank Philip Schütte for his contributions to and critical review of this chapter. The research project NamiRo was funded by the German Federal Ministry of Education and Research (BMBF) within the funding priority Social-Ecological Research (SÖF) and the funding programme Sustainable Economy (Nachhaltiges Wirtschaften). The Standards and the Extractive Economy Report was funded by the Intergovernmental Forum on Mining, Minerals, Metals and Sustainable Development.

References

Alliance for Responsible Mining. (2018). *CRAFT—Code of Risk-mitigation for artisanal and small-scale mining engaging in Formal Trade—Version 1.0.* https://www.responsiblemines.org/wp-content/uploads/2018/08/2018-07-31-CRAFT-Code-v-1.0-EN.pdf. (Accessed 10 July 2019).

Alliance for Responsible Mining. (2019). *Fairmined achieves selling of 1 t of gold.* http://fairmined.org/ton-fairmined-gold/. (Accessed 6 July 2019).

Amnesty International Ltd. (2016). *This is what we die for: Human rights abuses in the Democratic Republic of the Congo power the global trade in cobalt.* https://www.amnesty.org/en/documents/afr62/3183/2016/en/. (Accessed 5 July 2019).

BGR—Bundesanstalt für Geowissenschaften und Rohstoffe. (2017). *Vorkommen und Produktion mineralischer Rohstoffe—ein Ländervergleich.* Hannover, (in German).

BGR—Bundesanstalt für Geowissenschaften und Rohstoffe. (2019). *Raw materials information system.* Hannover, not public.

CCCMC—China Chamber of Commerce of Metals, Minerals and Chemicals Importers and Exporters. (2016). *Chinese due diligence guidelines for responsible mineral supply chains.* http://www.cccmc.org.cn/docs/2016-05/20160503161408153738.pdf. (Accessed 20 April 2019).

Cook, R., & Mitchell, P. (2014). *Evaluation of mining revenue streams and due diligence implementation costs along mineral supply chains in Rwanda.* Bundesanstalt für Geowissenschaften und Rohstoffe. https://www.bgr.bund.de/EN/Themen/Min_rohstoffe/ CTC/Downloads/. (Accessed 20 June 2019).

Divekar, A. (2018). *Vedanta Resources' anti-green image could impact the brand globally.* Business Standard. https://www.business-standard.com/article/companies/vedanta-resources-anti-green-image-may-hit-brand-scare-away-investors-118060600313_1. html. (Accessed 5 June 2019).

Ernst & Young. (2018). *10 business risks facing mining and metals.* https://www.ey.com/en_ gl/mining-metals/10-business-risks-facing -mining-and-metals. (Accessed 5 July 2019).

EU—European Union. (2017). Regulation (EU) 2017/821 of the European Parliament and of the Council of 17 May 2017 laying down supply chain due diligence obligations for Union importers of tin, tantalum and tungsten, their ores, and gold originating from conflict-affected and high-risk areas. *Official Journal of the European Union, 130.* Bruessels.

Gronwald, V. (2019). *Der Goldsektor in Deutschland—Marktstudie für verantwortungsvolles Gold aus dem Kleinbergbau.* Hannover: Bundesanstalt für Geowissenschaften und Rohstoffe [in German].

Hilson, G., Hilson, A., & Mc Quilken, J. (2016). Ethical minerals: Fairer trade for whom? *Resources Policy, 49,* 232–247.

IPIS—International Peace Information Service. (2019). *Maps of DRC, Antwerp.* http://ipis-research.be/home/conflict-mapping/maps/conflict-mapping-drc/. (Accessed 22 May 2019).

ISEAL Alliance. (2016). *Chain of custody models and definitions.* https://www.isealalliance. org/sites/default/files/resource/2017-11/ISEAL_Chain_of_Custody_Models_Guidance_ September_2016.pdf. (Accessed 15 July 2019).

Kickler, K., & Franken, G. (2017). *Sustainability schemes for mineral resources: A comparative overview* (p.167). Hannover: Bundesanstalt für Geowissenschaften und Rohstoffe.

Kühnel, K., Schütte, P., Bach, V., Franken, G., Dorner, U., & Finkbeiner, M. (2019). *Promoting the social license to operate through integrating social and environmental aspects in criticality assessments.* In *Conference proceedings, 9th international conference sustainable development in the minerals industry, Sydney, 27–29 May 2019, Australia*: The Australasian Institute of Mining and Metallurgy. Publication Series No 2/2019.

Maennling, N., & Toledo, P. (2018). *The Renewable Power of the Mine.* Columbia Centre on Sustainable Investment. http://www.bmz.de/rue/includes/downloads/CCSI_2018_-_The_ Renewable_Power_of_The_Mine__mr_.pdf. (Accessed 15 July 2019).

Marscheider-Weidemann, F., Langkau, S., Hummen, T., Erdmann, L., Tercero Espinoza, L., Angerer, G., et al. (2016). *Rohstoffe für Zukunftstechnologien 2016.* DERA Rohstoffinformationen 28, Berlin.

Matthysen, K., Spittaels, S., & Schouten, P. (2019). *Mapping artisanal mining areas and mineral supply chains in Eastern DR Congo, International Peace Information Service (IPIS) and Danish Institute for International Studies (DIIS), Antwerp.* http://ipisresearch.be/wp-content/uploads/2019/04/1904-IOM-mapping-eastern-DRC.pdf. (Accessed 7 July 2019).

McManus, M. C. (2012). Environmental consequences of the use of batteries in low carbon systems: The impact of battery production. *Applied Energy, 93*, 288–295.

Munk, L. A., Hynek, S. A., Bradley, D. C., Boutt, D., Labay, K., & Jochens, H. (2016). Lithium brines: A global perspective. *Reviews in Economic Geology, 18*, 339–365.

Neumann, M., Barume, B., Ducellier, B., Ombeni, A., Näher, U., Schütte, P., et al. (2019). *Traceability in artisanal gold supply chains in the democratic republic of the Congo.* Hannover: Bundesanstalt für Geowissenschaften und Rohstoffe.

Oberle, B., Bringezu, S., Hatfeld-Dodds, S., Hellweg, S., Schandl, H., Clement, J., et al. (2019). *Global resources outlook 2019: Natural resources for the future we want.* Nairobi: United Nations Environment Programme.

OECD—Organisation for Economic Co-Operation and Development. (2016). *OECD due diligence guidance for responsible supply chains of minerals from conflict-affected and high-risk areas* (3rd ed., p. 118). Paris: OECD Publishing. https://.doi.org/10.1787/9789264252479-en. (Accessed 5 April 2019).

Potts, J., Wenban-Smith, M., Turley, L., & Lynch, M. (2018). *Standards and the extractive economy.* International Institute for Sustainable Development, ISBN: 978-1-894784-79-5. https://www.iisd.org/sites/default/files/publications/igf-ssi-review-extractive-economy.pdf. (Accessed 15 July 2019).

PRI—Principles for Responsible Investment. (2019). *Vale dam collapse reignites mining safety pressure from investors.* https://www.unpri.org/pri-blog/vale-dam-collapse-reignites-mining-safety-pressure-from-investors-/4293.article. (Accessed 4 July 2019).

RMI—Responsible Minerals Initiative. (2018). *Blockchain guidelines.* http://www.responsiblemineralsinitiative.org/media/docs/RMI%20Blockchain%20Guidelines%20-%20FINAL%20-%2012%20December%202018.pdf. (Accessed 15 July 2019).

RMI—Responsible Minerals Initiative. (2019). *Responsible Minerals Assurance Process.* http://www.responsiblemineralsinitiative.org/responsible-minerals-assurance-process. (Accessed 17 May 2019).

Schütte, P. (2019). International mineral trade on the background of due diligence regulation: A case study of tantalum and tin supply chains from East and Central Africa. *Resources Policy, 62*, 674–689.

Seccatore, J., Veiga, M., Origliasso, C., Marin, T., & De Tomi, G. (2014). An estimation of the artisanal small-scale production of gold in the world. *Science of the Total Environment, 496*, 662–667.

UN—United Nations. (2019). *Letter dated 6 June 2019 from the group of experts on the Democratic Republic of the Congo addressed to the President of the Security Council.* https://undocs.org/S/2019/469. (Accessed 18 June 2019).

Vidal, O., Goffé, B., & Arndt, N. (2013). Metals for a low-carbon society. *Nature Geoscience, 6*, 894–896.

Voora, V., Bermudez, S., & Larrea, C. (2019). Global market report: Coffee. *Sustainable commodities marketplace series.* State of Sustainability Initiatives. https://www.iisd.org/sites/default/files/publications/ssi-global-market-report-coffee.pdf. (Accessed 15 July 2019).

World Bank. (2019). *Climate-Smart Mining: Minerals for Climate Action, Infographic.* http://www.worldbank.org/en/news/infographic/2019/02/26/climate-smart-mining. (Accessed 20 June 2019).

World Gold Council. (2019). *Goldhub, gold supply and demand statistics.* https://www.gold.org/goldhub/data/gold-supply-and-demand-statistics. (Accessed 2 July 2019).

The role of a circular economy for energy transition

12

James R.J. Goddin

Granta Design, Rustat House, Cambridge, United Kingdom

Introduction

Most of today's technologies depend on alloys comprised of collections of elements each with its own complex international supply chain. Many of the elements in use today have seen comparatively little use until relatively recently. In the case of telecommunications, which used to use only a handful of elements from the periodic table (primarily copper), we now use most of the commercially exploitable elements that exist (see Wang et al. within this book).

Renewable and low carbon technologies are no different in this respect. The technologies used in photovoltaic materials currently use various combinations of silicon, gallium, arsenic, tellurium, cadmium, copper, indium, and selenium, amongst others, in order to provide the technical properties that allow them to operate efficiently (see Wang et al. and Zepf within this book). For full/hybrid electric vehicles and wind turbines, there is a heavy reliance upon neodymium, praseodymium, dysprosium, samarium, cobalt, and copper for electric motors/generators and a strong reliance upon lithium and graphite for energy storage (see Weil et al. within this book).

While there is no absolute scarcity of any of these elements, their production is, however, restricted by their geographical concentration, the economics of producing them at scale, and the rate at which high-volume demand has emerged.

The production concentration of materials is most commonly described by the Herfindahl-Hirschman Index (HHI), which describes the level of monopoly as a function of the number of producers and their relative output. An HHI score approaching zero indicates an evenly distributed supply chain with little market dominance, a score approaching 10,000 indicating a completely monopolistic supply chain.

Fig. 12.1 highlights the trends in HHI for the elements commonly used in electric and hybrid vehicles and illustrates that while well-established bulk materials such as copper may be relatively nonmonopolistic, the rare-earth elements neodymium, dysprosium, and samarium have all been highly monopolistic for at least the past 20 years. The supply of these elements is dominated by Chinese production and this has been a focal point for trade disputes and a case brought against China through the WTO as a result of export restrictions (WTO, 2014).

The Material Basis of Energy Transitions. https://doi.org/10.1016/B978-0-12-819534-5.00012-X

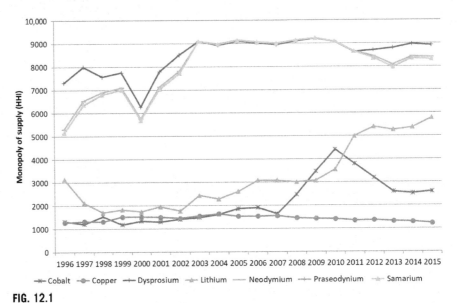

FIG. 12.1

Monopoly of supply of key elements used in electric vehicles.

Credit: Own figure based on production data from the USGS and BGS.

Fig. 12.1 also illustrates that lithium has become steadily more monopolistic since the turn of the millennium, supply in this case being increasingly dominated by Australia. Cobalt experienced a period of increased monopoly between 2007 and 2010 due to increased production by the Democratic Republic of Congo and has since recovered, partially due a decrease in supply from this region and also due to modest additional production from alternative regions.

Monopoly of supply is just one indicator and it is common for HHI to be modified using other, country-specific indicators (such as geopolitical stability or environmental maturity) to reflect different types of risk that are country specific and which might impact supply and ultimately be reflected in the price volatility of the resource. For a critical reflection, see Paul Gilbert within this book. Fig. 12.2 reflects two modifications of the HHI taking into account the World Bank Governance Index as a metric of "Sourcing and Geopolitical Risk" and Yale's Environmental Performance Index as a metric of "Environmental Country Risk" (i.e., the environmental maturity of the producing regions).

Fig. 12.2 highlights the perceived relative geopolitical stability and environmental maturity of the supply chains for the key elements used in hybrid/electric vehicles and particularly highlights the rare-earth elements (samarium, dysprosium, neodymium, and praseodymium) and to a lesser extent cobalt. The figure thereby serving to suggest that the monopoly observed in Fig. 12.1 may be of particular concern for these reasons.

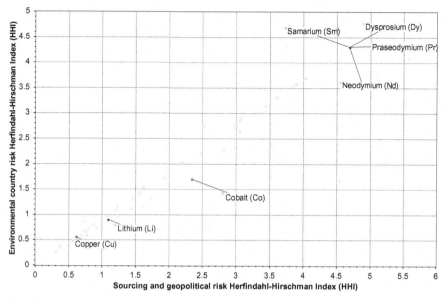

FIG. 12.2

Modified HHI indexes for the commercially produced elements in the periodic table highlighting those used in hybrid/electric vehicles, reflecting sociopolitical and environmental governance risks associated with production.

Credit: ANSYS Granta, CES EduPack.

Fig. 12.3 highlights the embodied energy of production[a] for the key elements used in hybrid/electric vehicles and their recent price volatility [calculated as the percentage difference between the maximum and minimum price (USD/kg) over the past 5 years, relative to the minimum price]. Embodied energy of production is the energy consumed by all of the processes associated with the production of a product, from the mining and processing of natural resources to manufacturing (see Penaherrera and Pehlken within this book). This figure serves to highlight a possible manifestation of the risks indicated in Figs. 12.1 and 12.2 through the resulting price volatility observed for the rare-earth elements and cobalt. When combined with the previous figures, it also highlights that elements that have the largest embodied energy per kg are also dominated in their production by countries with a lower overall environmental maturity, which should be of concern for low carbon technologies employing these materials.

Monopolistic supply chains are, generally speaking, more likely to experience price volatility as the likelihood of a supply disruption impacting the availability of the material is greater. Monopolistic supply chains are also more likely to exist for elements that have only started to see commercial use more recently and, therefore, may also be indicative of a lower overall market maturity. It also tends to be the case that production of materials subject to monopolistic supply tends to manifest in countries with lower

---

[a]The energy required to make 1 kg of the material from its ores or feedstocks.

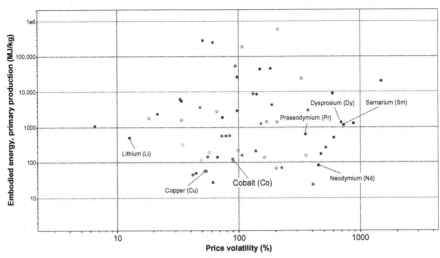

FIG. 12.3

The embodied energy and price volatility of the key elements used in hybrid/electric vehicles.

Credit: ANSYS Granta CES EduPack.

environmental maturity and/or sociopolitical accountability and may, therefore, carry with them disproportionate environmental, ecological, or social impacts.

In the case of technologies where rapid adoption and expansion is being promoted, a monopolistic supply chain is also less likely to scale as readily to meet demand than a nonmonopolistic one due to relative scalabilities of production at single or multiple production sites.

Resource supply and competition

Why then are such materials subject to monopolistic supply when there is a clear and growing market demand? Mining is a highly capital-intensive investment, and one that might take many years to become economically self-sustaining. In the 2009 rare earths crisis, which impacted amongst other elements, neodymium, dysprosium, praseodymium, and samarium, one of the responses was the reopening of the Mountain Pass mine in the United States. The intention of this investment being to create an alternative source, which would benefit from the increased prices of the time and significantly reduce the strong Chinese monopoly. Unfortunately, the mine was only beginning to reach production by the time China lost the WTO case brought against it and the subsequent release of significant volumes of inexpensive rare earths onto the market, at prices below what was economically viable for Mountain Pass, resulted in the owners filing for bankruptcy. This in turn led to the retention of the Chinese monopoly for the production of these materials.

There are many reasons why China is able to produce rare-earth metals at a lower price than Western mines. First, they have a long-term national strategy for the use and development of these materials and the products produced from them. Second, there has reportedly been a significant level of illegal mining and indeed closing down illegal mines was one of the justifications given by China for reducing exports in 2009. Third, the costs for producing the materials are lower, in part due to labor costs, in part due to extensive state investment in expertise and technology for extracting and refining the materials. Some would also argue that costs are kept lower through a lower overall adherence to environmental standards, particularly in the case of illegal mines and refineries. To maintain economic viability outside China, a competing nation would have to sustain production based on a product that would be more expensive to produce than the Chinese alternative and, working on established trade and financing rules, this is not easy to achieve outside the application of non-trade barriers or artificially introduced trade tariffs.

Some minerals, such as tungsten, tin, tantalum, and gold (3TG) are subject to regulations seeking to disconnect procurement of materials, which have been used to fund conflict. One example of this is the US Dodd-Frank Act on conflict minerals and there will be a similar EU Conflict Minerals legislation coming into effect in 2021 (see Franken et al. within this book). Price volatility for 3TG may be expected, linked to the additional administrative burden of complying with this legislation and a temporary shift to known conflict-free sources, while certified conflict-free refineries are established.

Reducing supply risks using circular economy

One possible partial solution for companies reliant upon materials at risk of supply disruption is to reduce their dependence on new/virgin materials through the application of circular economy principles—including reuse, remanufacturing and recycling as well as enhanced durability and utility of materials. Indeed, the economic benefits of this approach may be quite significant if properly implemented.

Fig. 12.4 is commonly referred to as the "butterfly diagram" and illustrates the material flows of a circular economy. The main principles of the circular economy being the use of renewable energy, the separation of technical and biological material flows (to avoid contamination) and a focus on maintaining materials and products in use and at their highest level of economic value (i.e., products > parts > materials).

The circular economy thereby considers maintaining not just the value of the materials used in the manufacture of the product but also the added value from manufacturing parts and assembling products, intellectual property, and marketing value created by the producer of the product. A convenient illustration of the benefits of this is given by considering the value of the materials used in the production of an iPhone, a mere $1.03 (Von Kessel, 2017). If we compare this with the much higher retail value of the product (hundreds of dollars), then the added value from investment in technologies required to make the product work, the manufacturing investment and

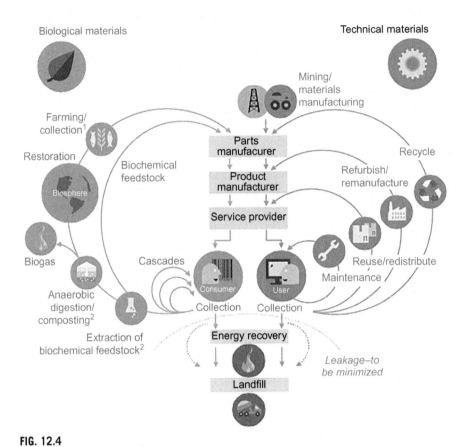

FIG. 12.4

The "butterfly diagram" for the circular economy from the Ellen MacArthur Foundation.

Credit: Ellen MacArthur Foundation (2020). Infographic. https://www.ellenmacarthurfoundation.org/circular-economy/concept/infographic. Accessed 17 March 2020.

supply chain infrastructure required to produce the product at volume, the software, supply networks, retail outlets, and branding become apparent. By focusing on the reuse of products and parts, much of this additional value may be recovered and not just the much lower raw material value.

By maintaining products, parts and materials in these loops for longer, the need for new materials may be substantially reduced and the same overall capacity realized for a considerably smaller materials footprint. Likewise, by maintaining products in service for longer or by utilizing products more heavily, the need for additional products to provide capacity at a systems level is reduced—decreasing the overall material requirement further still. This approach is often manifested within what has become known as the "sharing economy," where products are leased or provided as a service rather than owned (Ellen MacArthur Foundation, 2019a).

Voluntary retained ownership of products through the sharing economy or involuntary retained ownership through extended producer responsibility schemes (OECD, 2001) brings with it a shift in design and engineering mindset. Rather than producing products for the lowest production cost, there becomes a strong incentive to accrue value and minimize costs by keeping the product in service for longer, leading to a focus on longevity/durability. Likewise, there becomes a strong incentive to ensure the costs of ultimately recovering the value in parts and materials at the end of use are maximized by employing philosophies such as design for disassembly, repairability, modularity, and upgradability—practices that are otherwise frequently ignored by conventional "linear" product models. Shifting to a circular economy can also require carefully considered changes to supply chains and supply chain relationships. For example, many jobs rely on producing new materials and on manufacturing these into products, without close consultation and the realignment of effort within the supply chain toward a new circular economy model, these jobs may be disrupted and the economic, environmental and social benefits may not be realized.

The circularity of a product can be considered simply as a function of:

1. The flow of materials into the product—how much comes from reuse or remanufacturing, and how much recycled content is used rather than new virgin materials.
2. The material flows from the product—how much is reused, remanufactured, or recycled rather than being landfilled, incinerated, or downcycled at the end of life.
3. How long and how intensively the product is used.

The relationship between these factors and the "circularity" of a product have been described by the Ellen MacArthur Foundation and Granta Design in the Materials Circularity Indicators methodology, which evaluates the circularity of products, portfolios, and companies as a result of these factors (Ellen MacArthur Foundation and Granta Design, 2015).

While a pursuit of circularity is to be encouraged from a resource efficiency perspective, it is the benefits that accrue to the supply chains involved that are perhaps most of interest and are the motivation for most businesses. As the term "circular economy" indicates, the primary benefit of interest is the economic return from the circular model—this can be direct economic returns through circularity itself, extracting additional value directly, it can also be an indirect economic gain through reducing exposure to price volatility through sources we have discussed earlier in this section.

There are also often significant environmental benefits that accompany the circular economy and a well-rounded approach will consider these as well as part of optimizing the product model. The potential environmental benefits from the circular economy have been amply illustrated by the recent works (Ellen MacArthur Foundation, 2019b).

Emerging examples of the circular economy can already be observed within the low carbon vehicles market. As most vehicles spend 92% of their time parked (i.e., not in use), a shift to access-based models or autonomous driving offer potential solutions to an otherwise underutilized resource. By enabling access-based models, vehicles can be used on demand rather than owned. In the case of Tesla's vision, vehicles can even be bought and then released by their owners to generate income from autonomously driving others when not in use (Musk, 2019). Such models may in future substantially reduce the overall number of vehicles on the road, meaning that the quantity of materials required to move a population may become substantially reduced.

Such access-based models may of course place a much heavier load on the product. Under traditional ownership, a vehicle with an ultimate lifetime mileage of 300 kmiles and an owner who drives 30 kmiles per year might be expected to last 10 years. The same vehicle being utilized for the additional 92% of its available time would perhaps last less than a year. A necessary focus on the durability of electric vehicles, particularly the valuable electric motors and batteries, is obvious. Tesla's stated aim to deliver a million-mile vehicle (Oberhaus, 2019) would substantially increase the utility of vehicles even when used under such a shared ownership or access-based model.

Interestingly, shifting to such a model may have a significant impact on the sale of competing vehicles and hence on market dominance—if the needs of 12 people can be serviced through one vehicle, those people may no longer be inclined to purchase the other 11 vehicles they would have otherwise bought from the market. This leads to a disproportionate market dominance compared with the volume of product sold and this is perhaps a new commercial paradigm that will accompany the emergence of the circular economy. Even with greater longevity and access-based models, there is still an ultimate need to separate products, components, and materials for reuse, remanufacturing, and recycling at the end of their serviceable lifetime, and this is an area that remains fairly immature for many renewable or low carbon technologies.

Product design and materials recovery

Permanent magnets in hybrid/electric vehicles and in wind turbines are commonly permanently bonded within the motors and generators—delivering reliability and performance but at the expense of recoverability. Efforts to deliver both reliability and recoverability remain primarily at the prototype stage with one of the best examples being the DEMETER prototype recyclable motor (DEMETER, 2019).

One of the characteristics of the prevalent linear economy is that economic growth depends heavily upon the extraction of resources (if these resources are not grown, then they are mined) the conversion of resources into products and the disposal of these products back into the environment. In times where resources were cheap and plentiful and the impacts from energy consumption and waste were of little concern, this linear "take-make-dispose" model served well and resulted in the inexpensive disposable products we have come to enjoy and often depend on in the Western world.

Those times have, however, come to an abrupt end and consumers are also increasingly expecting the delivery of low carbon technologies—leading to rapidly growing demand for materials for renewable technologies at a time when yields from mining activities are at an all-time low. The circular economy offers one potential so-lution if implemented correctly but it is clear that recycling alone will not be enough.

As mentioned earlier, the circular economy offers only a "partial" solution to the problems of access to materials, this is certainly the case where there is market expan-sion. For any technology with growing adoption, there will continue to exist a strong demand for new or virgin material even when reuse or recycling of used products is widespread. This is because the balance of demand that cannot be fulfilled through the circular economy needs to be satisfied through primary production. An inspection of most commercial metals in use today will indicate that, despite often very high re-cycling rates, primary production has continued to expand significantly, maintaining a highly linear model despite a relatively circular material recovery process.

In order to deliver widespread recoverability of valuable resources at a time of market expansion and a widespread shift to new technologies, the recovery of ma-terials needs to be taken to a completely different level. We are already seeing the widespread implementation of extended producer responsibility for products, putting the onus on manufacturers for the recovery of materials from the products they sell onto the market (DEFRA, 2019).

There is also a growing focus on the circular bioeconomy, seeking to replace techni-cal material flows with biological flows, some great examples of the circular bioeconomy are seen in the work of Rotorua-based Scion (Scion, 2019). While this is unlikely to displace the use of materials used today in batteries, magnets, or electrical systems, the circular bioeconomy is likely to see greater uptake for other more structural components, where the ability to naturally replenish the materials used has many obvious advantages.

A further extension to the circular economy that is starting to be considered by some is the standardization of materials and designs across vendor platforms. Whilst such a model is undoubtedly difficult to achieve from the standpoint of intellectual property and commercial propriety, such an approach might yield the best overall use of materials by enabling end of life products to be more readily utilized and recovered within a specific sector, keeping products at their highest value for longer, minimizing end of life costs through economies of scale, and more effectively decou-pling the products from primary resource use. Examples of this approach are starting to emerge for packaging systems (Loop, 2019).

Other transitions are likely to be more societal and some may even revolve around the reevaluation of our own expectations as consumers—indeed there are early signs that this is already happening. While much of the growth in demand for materials today is driven by a large emerging middle class, the reason we consume so much material is also often not driven by necessity but rather by fashion and our own desire to express economic and social success through our consumption (see Sonnberger within this book). This is apparent through the widespread replacement of technolo-gies that have not failed technically but are put aside once newer alternatives are available. This is a far more deep-seated challenge and one that requires a fundamen-tal reevaluation of the measures of personal success.

Early examples of this type of social transition include a growing demand for access over ownership, for experiences over acquisitions, for work flexibility over job security. In other cases, it is reflected by an increased willingness by consumers to spend more on products and services that align with their personal values on the environment and social and ecological good, such social changes being more apparent for Millennials (Neilsen, 2018).

Such social changes are more apparent for Millennials and Gen. Z and have already led to some very significant businesses, such as Unilever, realigning their business models, and brand propositions with these new aspirations (BBC, 2019). It is also leading the emergence of completely new businesses, such as RiverSimple set on competing with established brands. Even in the finance sector, there are stark warnings for business to take change more seriously, with the governors of the Bank of England and Bank of France warning that "If some companies and industries fail to adjust to this new world, they will fail to exist" either because ethically minded investors will place their money elsewhere, or the market for their products will disappear completely (Bank of England, 2019). It is likely that for some sectors this will lead to stranded assets—i.e., investments in infrastructure that will never generate enough income to pay off the money invested (Chapman, 2019).

Such societal shifts are also acknowledged to offer significant opportunities to brands that do adapt. In many service-based models, there are opportunities to increase brand loyalty and consumer engagement through more frequent contact with the consumer, while offering them greater freedom and flexibility over the products and services they use. How exactly this more intrinsic, value-driven change will evolve in the long-term remains to be seen.

Ultimately, the reduction of the environmental and social impacts from material consumption, even with a strong circular economy, will require society to reduce the overall levels of material demand to the point that a far greater portion of what we use can originate from the resources that we have already extracted. For the materials that we continue to extract, a level playing field will need to be established with recycled or reused materials, this may mean paying more for virgin materials—something that may happen anyway as the costs of extraction continue to rise.

The costs of materials we consume today do not commonly cover the full economic, environmental, and societal costs of production and indeed there is probably a close alignment ultimately between the circular economy in its final state and various fair-trade initiatives. There remains an open question, however, on how fair trade will be applied to the circular economy, should there perhaps be a form of "circular fair trade," where some of the long-term economic value realized through the recirculation of materials vests back to supporting the environments, ecosystems, and societies that were impacted by the original extraction?

Conclusions

The commercial application of circular economy efforts, such as outlined in this chapter, will no doubt become essential if effective scale up of renewable technologies is to become economically and environmentally sustainable in the long term, otherwise the continued extraction of virgin raw materials will continue and these valuable resources will eventually be subject to significant price increases as ore grades decrease and extraction costs rise. This may eventually lead to the limitation of growth (or at least replenishment) of renewable technologies and lead society away from the low carbon future that is so widely recognized as being imperative.

References

Bank of England. (2019). https://www.bankofengland.co.uk/news/2019/april/open-letter-on-climate-related-financial-risks. (Accessed November 27, 2019).

BBC. (2019). https://www.bbc.com/news/business-49923460. (Accessed November 27, 2019).

Chapman, B. (2019). *Mark Carney tells banks they can't ignore climate change.* 17 April 2019. The Independent.

DEFRA. (2019). *Reforming the UK packaging producer responsibility system: Summary of responses and next steps.* https://assets.publishing.service.gov.uk/government/uploads/system/uploads/attachment_data/file/819467/epr-consult-sum-resp.pdf. (Accessed November 27, 2019).

DEMETER. (2019). *European training network for the design and recycling of rare-earth permanent magnet motors and generators in hybrid and full electric vehicles* EU Horizon 2020 Project. Grant Agreement no 674973; 'New video released on first recyclable e-motor (EUDEMETER)'. https://etn-demeter.eu/new-video-first-recyclable-e-motor/. (Accessed November 27, 2019).

Ellen MacArthur Foundation. (2019a). *Amsterdam sharing economy action plan.* https://www.ellenmacarthurfoundation.org/case-studies/. (Accessed November 27, 2019).

Ellen MacArthur Foundation. (2019b). *Completing the picture: How the circular economy tackles climate change.* https://www.ellenmacarthurfoundation.org/assets/downloads/Completing_The_Picture_How_The_Circular_Economy-_Tackles_Climate_Change_V3_26_September.pdf. (Accessed November 27, 2019).

Ellen MacArthur Foundation, Granta Design. (2015). *Material circularity indicators methodology.* https://www.ellenmacarthurfoundation.org/assets/downloads/insight/Circularity-Indicators_Project-Overview_May2015.pdf. (Accessed November 27, 2019).

Loop. (2019). https://loopstore.com/. (Accessed November 27, 2019).

Musk, E., 2019. Twitter: @elonmusk. April 2019.

Neilsen. (2018). *Global consumers seek companies that care about environmental issues.* https://www.nielsen.com/eu/en/insights/article/2018/global-consumers-seek-companies-that-care-about-environmental-issues/. (Accessed November 27, 2019).

Oberhaus, D. (2019). *Tesla may soon have a battery that can last a million miles.* 23 September 2019Wired.

OECD. (2001). *Extended producer responsibility—A guidance manual for governments.* https://doi.org/10.1787/9789264189867-en.

Scion. (2019). https://www.scionresearch.com/home. (Accessed November 27, 2019).

Von Kessel, I. (2017). *The materials that make up the iPhone.* https://www.statista.com/chart/10719/materials-used-in-iphone-6/. (Accessed November 27, 2019).

WTO. (2014). *DS431: China—Measures related to the exportation of rare earths, tungsten and molybdenum.* https://www.wto.org/english/tratop_e/dispu_e/431_432_433r_e.pdf. (Accessed November 27, 2019).

Substitution of critical materials, a strategy to deal with the material needs of the energy transition?

13

James R.J. Goddin

Granta Design, Rustat House, Cambridge, United Kingdom

Material substitution for renewable energy technologies

Practical limitations are emerging to the scalability of some renewable technologies these primarily relate to the challenges of scaling the extraction and production of mineral resources needed to manufacture these technologies at scale and fast enough to keep up with the magnitude of the problem presented by long-term global inaction on climate change.

This chapter discusses what opportunities exist to "side-step" these restrictions through substitution.

In exploring this question, is it perhaps useful to use, as a construct, the different types of substitution that are possible—as aptly outlined by the CRM_InnoNet project, namely (CRM_InnoNet, 2019):

1. Substance for Substance Substitution
2. Service for Product Substitution
3. Process for Process Substitution
4. New Technology for Substance Substitution

The following chapters will discuss these different types of substitution in more detail.

Substance for substance substitution

Low carbon technologies typically utilize a suite of elements from the periodic table to provide the technical performance required (see Zepf, Wang et al., and Goddin within this book). For motors in hybrid/electric vehicles and generators in wind turbines, for example, the necessary performance required is the ability of a magnetic field to convert energy to motion or motion to energy.

Permanent magnets require a favorable remnant induction (field strength per unit area). The materials selected must also typically have a high curie temperature (the

The Material Basis of Energy Transitions. https://doi.org/10.1016/B978-0-12-819534-5.00013-1

temperature at which magnetization is lost) as the magnets are usually exposed to elevated temperatures when in use.

For hybrid/electric vehicles, there is also a strong incentive to minimize the weight to maximize range, and therefore the field strength per unit mass becomes important. For both vehicle and wind turbine applications, corrosion resistance is another common attribute due to the salinity of the environment to which the product may be exposed—although this is often solved through the application of coatings or by encapsulating the magnets to avoid exposure.

While supply constraints often apply to specific elements of the composition of materials, it is not the element that we are seeking to substitute but the performance of the material that is enabled by that element—i.e., the net effect of all of the elements in the composition and what these, acting together, will allow the material to do (see Fig. 13.1).

It is an unfortunate reality that the more constraints we apply to a material selection process, the fewer materials will typically match our requirements and the more compromises we are likely to have to make if we are seeking to identify a substitute material for one that is subject to supply constraints. This is undoubtedly the case at present with permanent magnets used in low carbon applications, where the rare-earth elements (typically neodymium, praseodymium, dysprosium, and samarium) within them are responsible for the high remnant induction and high curie temperatures required. Trying to reduce supply risks by removing these elements often leads to significant losses in the properties available and compromises one or more aspects of the application itself.

One possible solution is to substitute the rare-earth elements used in magnets for other rare-earth elements that are more abundant. The rare-earth group comprises 17 elements that are typically found in combination with each other in mixed rare-earth oxides and are refined through sequential separation, in general terms by removing the lighter elements from the heavier elements sequentially. Fig. 13.2 highlights the relative abundance of these 17 elements in the Earth's crust and the annual world production of these elements.

The heavy rare earths that provide the high-temperature performance of magnets (typically dysprosium and praseodymium) are often only used in trace quantities, they are, however, amongst the least abundant and are more expensive to refine, being some of the last elements separated from mixed rare-earth oxide. Neodymium typically makes up a more substantial part of neodymium-iron-boron permanent magnets just as samarium tends to comprise a more significant proportion of samarium-cobalt-based magnets, both are more abundant and produced in higher volumes and from a greater variety of sources.

Cerium is, however, one of the first rare earths to be separated from rare-earth oxide and often comprises four times as much of the composition as neodymium. Cerium is, however, considered of lower value as supply can easily outstrip demand, and indeed the lack of market for cerium can in some cases significantly impact the production economics of a rare-earth mine, as it still has to be separated first, before the more valuable rare earths can be accessed.

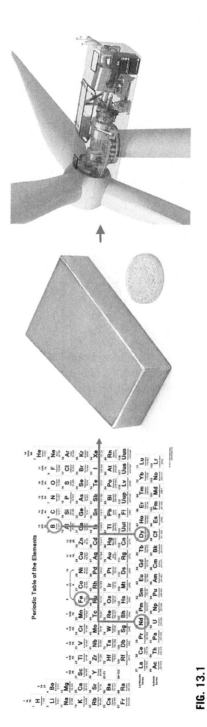

FIG. 13.1

Neodymium, dysprosium, iron, and boron provide specific functions in wind turbines.

Credit: Figure made by the author.

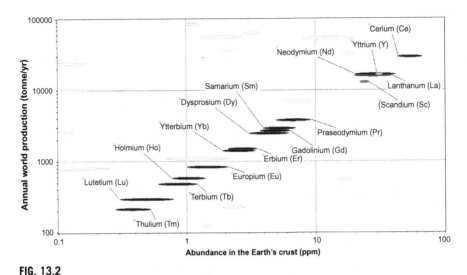

FIG. 13.2

Abundance in earth's crust vs. annual world production highlighting the rare-earth elements.

Credit: Granta CES EduPack.

A project led by the Ames National Laboratory in Iowa has been seeking to increase the demand for cerium by researching its use as a substitute for neodymium in magnet applications, if successful, this would have the effect of substantially increasing the utility of rare-earth elements in magnets and would also support the production economics of rare-earth separation more generally (AMES, 2011). This project is an excellent example of substance for substance substitution. However, the costs and time frames involved in identifying and commercially implementing such substitutions should not be underestimated, as discussed later.

Service for product substitution

The notion of substituting a product with a service is a central theme behind many circular economy models. At its root, is the realization that someone who buys a product sometimes does not want the product itself but instead the function the product serves—one of the most common analogies being a consumer not wanting to own a drill but wanting to have the ability to make holes.

This realization of the desire for function rather than ownership brings with it a different paradigm, whereby a single product provides a service to a larger group of users rather than each user buying a separate product (and letting it sit idle for most of the time).

If implemented correctly, this form of substitution can lead to substantial savings for the users of the service (as hiring is typically cheaper than buying), it can lead to a higher level of personal freedom (as there is no product for the users to store or maintain), and it can lead to access to more top quality goods (as products designed to be hired out are often more durable and perform more reliably). For the service provider, it can also lead to increased revenues (as a smaller quantity of product can be leased many times over with a greater prospect of return business).

Traditional power generation is already an example of this type of service, after all, we do not typically go and buy a power station when we need power for our homes but buy the electricity produced by them from a service provider, often through some form of market intermediary.

Low carbon power generation is, however, more distributed in its nature and brings with it different types of service-based models, which include the traditional energy providers but also include "democratized services" such as local and community-level grids, where domestic power might just as easily come from your neighbors roof as from a commercial operator, household energy storage and vehicle-to-grid systems which are also starting to play essential roles to help overcome the more significant variation in supply inherent in many renewable energy systems.

Process for process substitution

Process for process substitution involves overcoming the need for a material in the first place by delivering the solution by a different means. An example of this is the delivery of wind power generation without rare-earth-based permanent magnets. Indeed this is where many of the earlier wind turbine designs started, and there are many in use today, utilizing gearbox-based designs instead with less expensive (and more abundant) magnetic materials (Morris, 2011).

There are, however, advantages to having a gearbox-free design, especially for offshore applications, where the cost of downtime and maintenance is significantly higher—gearboxes in wind turbines being a traditional weak point and a common source of failure.

Process for process substitution is commonly (but not always) accompanied by compromises in performance or means of deployment or maintenance regimes, quite often this type of change is also accomplished by reverting to an older solution rather than by moving to a new kind of technology due to the need for qualifying/certifying the technology, while keeping deployment risks to a minimum.

New technology for substance

An example of using new technology to substitute for a substance is apparent in some electric vehicle designs. Again, rare-earth elements often enable lightweight, power-dense motors that are attractive for automotive applications. Alternatives do, however, exist that do not require the use of rare-earth-based permanent magnets, and indeed the majority of Tesla's use switched reluctance motors instead. Switched reluctance motors do not have the same supply risk profile as rare-earth-based motors, as they do not contain permanent magnets, they also do not carry the same risks of demagnetization and are typically lower in cost initially. They can, however, be slightly heavier, a bit less efficient at lower speeds and can be prone to overheating if not carefully designed. The enabling technology for the use of switch reluctance motors in electric vehicles was lightweight transistors that enable the current high-speed switching required without significantly adding to the vehicle weight.

The substitution has, at least in part, enabled Tesla to become arguably the leading premium brand in electric vehicles, blazing a trail that most well-established incumbents in the internal combustion engine market are still struggling to catch. It is interesting to note, however, that the limitations mentioned above have come to bear on the smaller Tesla Model 3, which has moved to a permanent magnet-based design, most likely due to size, weight, and range benefits when coupled to a smaller vehicle platform (Bakker, 2018).

Other examples of this type of substitution are found in photovoltaics, where expensive silicon-based solutions have been replaced by less costly organic solar cells. The trade-off, in this case, is based on the cost per kilowatt-hour installed, the organic photovoltaics often being less efficient in transforming light into electricity but this being more than offset by the larger surface area that can be bought for the same investment as a silicon-based solution (Orgil, 2018).

Creating substitutes: By engineering simulation

It is perhaps now apparent that there are multiple trade-offs to be made when considering substitution and different types of substitution. It is important to realize, however, that substitution often is not an easy option, quite often there are a small number of "off-the-shelf" materials that can provide the specific functionality needed for a particular set of criteria and the best option will quite often be the incumbent one that one is seeking to substitute. Developing new materials is also traditionally a costly and time-consuming process and is often prone to dead ends and failed results, it is perhaps necessary to reflect upon how long it has taken to develop the handful of materials that we commonly use today and the level of investment it has taken to demonstrate that they can be used effectively in their specific applications.

That said, modern techniques are increasingly being brought to bear on the discovery of new materials and the rate of deployment into effective commercial solutions. Two projects stand out in this respect, in the United States, the Materials Genome Initiative and in the EU, the Accelerated Metallurgy project (Accelerated Metallurgy Project, 2016; Materials Genome Initiative, 2019). Both have sought to apply a combination of computational materials simulation and high-throughput materials synthesis and testing to search through many thousands of potential new alloys, increasing the number of combinations that can be explored by several orders of magnitude and significantly increasing the likelihood of identifying substitutes.

The ability of these projects to advance materials development to this degree has primarily been enabled by the development of advanced simulation tools (including machine learning and genetic algorithms), the deployment of advanced manufacturing techniques such as additive manufacturing and the availability of shared data management systems to collate and maintain the pedigree of the resulting materials data.

When combined, these technologies yield the significant acceleration in materials screening needed to identify promising candidate compositions. However, to develop a commercially relevant material, the material needs to be taken further still by understanding the impact of microstructural development as a function of manufacturing processes and how this influences the ultimate properties of the product. Engineering simulation has also developed significantly to address this need and indeed has benefitted from the availability of well-structured materials data that is increasingly common as part of materials discovery.

The combination of these technologies, while impressive, yields a materials development time that is still of the order of years, which is an improvement on the decades that would have traditionally been required but still represents a significant investment.

Barriers to substitution

Having highlighted how difficult it could be to identify possible substitutes for materials, it is essential also to understand that this is often just the beginning of the process of getting the material to market in an acceptable form.

Alternative materials will not perform in isolation, but need to be embedded into new designs, tested and developed into products and again tested in the field to determine the cost, technical viability, and reliability.

Many industries are also highly regulated and driven by robust certification or compliance requirements. Overcoming such conditions can often take years, depending upon the industry, and substituting something that is known to work and has already been through this process is often highly unpalatable. As a result, the path of substitution is rarely taken unless driven by a significant technical advance from a new material that has become available or by a regulatory requirement such as the banning of one of the substances used.

Conclusions

Substitution is a widely discussed topic and one which can have significant benefits if implemented correctly. There are multiple types of substitution, however, the process of identifying substitutes is often accompanied by compromises to performance, cost, or reliability. If the substitution involves developing new materials, this can often be a substantial undertaking and take years and significant levels of sustained funding to achieve. Modern materials data management systems and materials and engineering simulation tools have substantially reduced the costs and time frames involved and are likely to continue to do so.

References

Accelerated Metallurgy Project. (2016). *EU project 7th framework program grant agreement no 263206.* https://cordis.europa.eu/project/rcn/99430/factsheet/en. (Accessed 27 November 2019).

AMES. (2011). *Cerium-based magnets, REACT program.* https://arpa-e.energy.gov/?q=slick-sheet-project/cerium-based-magnets. (Accessed 27 November 2019).

Bakker, S. (2018). *Tesla Model 3 Motor—Everything I've been able to learn about it (welcome to the machine).* https://cleantechnica.com/2018/03/11/tesla-model-3-motor-in-depth/. (Accessed 27 November 2019).

CRM_InnoNet. (2019). http://www.criticalrawmaterials.eu/project-summary/. (Accessed 27 November 2019).

Materials Genome Initiative. (2019). https://www.mgi.gov/. (Accessed 27 November 2019).

Morris, L. (2011). *Direct Drive vs. Gearbox: Progress on both fronts, power engineering.* https://www.power-eng.com/2011/03/01/direct-drive-vs-gearbox-progress-on-both-fronts/#gref. (Accessed 27 November 2019).

Orgil, K. (2018). *Comparison of organic and inorganic solar photovoltaic systems.* San Luis Obispo: California Polytechnic State University.

Renewable energy technologies and their implications for critical materials from a sociology of consumption perspective: The case of photovoltaic systems and electric vehicles

14

Marco Sonnberger

ZIRIUS—Research Center for Interdisciplinary Risk and Innovation Studies, University of Stuttgart, Stuttgart, Germany

Introduction

Around the world, 57 countries have established 100% renewable electricity targets (Renewable Energy Policy Network, 2018). In this context, rooftop photovoltaic (PV) systems are the core renewable electricity technology that is suitable for integrating electricity production in private households. Thus, more and more private households are becoming electricity prosumers, both consuming and producing electricity (Ellsworth-Krebs & Reid, 2016). This is supported by government programs in many countries that involve citizens in the actual implementation of energy transitions. In 2017, electricity gained from renewables accounted for 26.5% of global electricity production (Renewable Energy Policy Network, 2018). Also in the context of mobility transitions, citizens are becoming an important success factor for the phaseout of petrol and diesel cars. While to date 11 countries have announced plans to phase out fossil fuel vehicles, many more have already set policy targets for the absolute number or overall share of electric vehicles (EVs) by a certain year (Partnership on Sustainable Low Carbon Transport, 2019). Accordingly, the worldwide stock of electric passenger vehicles has risen from less than 500,000 in 2013 to approximately 3 million in 2017 and is supposed to rise even more rapidly over the next decade (International Energy Agency, 2018). To date, however, only three countries—Norway, the Netherlands, and Sweden—have a passenger car stock share of electric cars of 1% or higher (International Energy Agency, 2018).

The Material Basis of Energy Transitions. https://doi.org/10.1016/B978-0-12-819534-5.00014-3

With the rise in renewable energy technologies and EVs, as well as the ongoing digitization of more and more areas of production and everyday life, concerns are increasingly being raised about the negative side effects and downsides of these developments. The growing use of critical materials in the respective technologies is one major point of criticism and concern. Sujatha Raman, for example, notes that renewable energy technologies are becoming fossilized, since the configurations of renewable energy systems increasingly resemble those of the old fossil energy regime, thereby reproducing hegemonic structures, inequalities, and environmental issues through the use of critical materials (Raman, 2013).

This chapter focuses on the role adopters of EVs and PV systems play with regard to critical materials embodied in these technologies. It aims to elaborate on why adopters of PV systems and EVs (can) hardly take into account the critical materials on which these technologies are based. My main argument will be that the adoption of PV systems and EVs is motivated by multiple factors, with environmental considerations being only one of many motives, and that, moreover, embodied critical materials lack discursive and cognitive salience and are thus ignored by most consumers.

The chapter is organized as follows: I start by defining the term consumption and pointing out the link between consumption and society-nature relations, in order to explain the co-constitution of consumption, production, and sustainability issues. In the following sections, I elaborate on the sociocultural and sociopsychological functions of consumer goods and draw on the two examples (adoption of EVs and PV systems) in order to discuss why consumers mostly overlook the use of critical materials embodied as invisible elements in renewable energy technologies. Finally, I conclude with a critical discussion of the role that consumers of PV systems and EVs play with regard to embodied critical materials.

Consumption and society-nature relations

Many scholars writing on consumption do not explicitly define the term. Thus, the term consumption often lacks clarity and is used in an arbitrary way (Evans, 2018). When scholars try to define consumption, they mostly conceptualize it as a process divided into different stages (Campbell, 1995; Evans, 2018; Fischer, Michelsen, Blättel-Mink, & Di Giulio, 2012; Schneider, 2000; Warde, 2005). Consumption is not a single act, but a sequence of diverse acts evolving over time (see Fig. 14.1).

The process of consumption starts with the selection of goods or services. This is followed by acquisition and use. The final stage of the process is disposal. In the

FIG. 14.1

Stages of the consumption process.

Credit: Own illustration.

selection phase, people search for information and engage in decision-making activities motivated by different kinds of purposes (e.g., utilitarian, expressive, contemplative, etc.), in order to acquire and use a specific service or good. The acquisition phase demonstrates the different ways in which people access goods or services. In the use phase, people integrate the acquired goods or services into their everyday lives, make use of them and—depending on the kind of good—literally consume them, thereby experiencing pleasure and satisfaction (in some cases also disappointment) from the usage. The disposal phase marks the point in the consumption process when people remove goods from their possession or abandon the use of services. With regard to goods, this does not automatically necessitate their wastage since many goods are disposed of without being broken, damaged, or used up (Evans, 2018).

It is obvious that the kind of goods and services people consume and the ways in which they are disposed of have significant socio-ecological consequences. Concerns about the unsustainability of consumption patterns have particularly been raised in the context of discussions around sustainable development. Hence, the depletion of critical materials embodied in different products and the issues related to recycling and reusing them after product disposal are some current debate topics concerning sustainable consumption. In the context of sustainable development, "sustainable consumption means that goods are acquired, used, and disposed of in such a fashion that all humans, now and in the future, are able to satisfy their (basic) needs and that their desire for a good life can be fulfilled" (Defila, Di Giulio, & Kaufmann-Hayoz, 2012: p. 13). This means that individual acts of consumption can be regarded as sustainable if they actually contribute to sustainable development. When assessing the sustainability of consumption activities, it is therefore crucial to differentiate between an intent-oriented and an impact-oriented perspective (Fischer et al., 2012; Stern, 2000). While people may have the intention to consume sustainably (intention-oriented perspective), this may not necessarily result in positive effects in terms of sustainability (impact-oriented perspective), and vice versa. Thus, there can be a discrepancy between environmental intent and environmental impact. For example, Stephanie Moser and Silke Kleinhückelkotten have shown that people who strongly self-identify as being environmentally aware use more energy and thus have a bigger carbon footprint than people who are less environmentally aware. Since both carbon footprint and environmental awareness are positively correlated with income, the authors conclude that the influence of environmental awareness is counterbalanced by an income effect (Moser & Kleinhückelkotten, 2018). Thus, environmental awareness does not necessarily translate into sustainable consumption patterns. The reasons for this are manifold and include, among others, motivational goal conflicts, contradictory and opposing behavioral incentives, as well as a lack of relevant knowledge, self-efficacy beliefs, and behavioral degrees of freedom (Kollmuss & Agyeman, 2002). Furthermore, criticizing and going beyond the focus on consumers as atomized decision-makers, others emphasize the embeddedness of individual behaviors in certain social and institutional contexts that lock consumers into unsustainable patterns of consumption (Sanne, 2002; Shove, 2010).

The socio-ecological impacts of consumption point to the fact that consumption patterns are a crucial element of so-called society-nature relations. The concept of society-nature relations describes the dynamic pattern of relationships between humans, society, and nature. Society-nature relations evolve from culturally specific and historically variable forms and practices through which individuals and groups of individuals at different scales (e.g., social milieus, nations, cultures) shape and regulate their relationships to nature (Becker, Hummel, & Jahn, 2011). Thus, society-nature relations "are concrete material relationships structured by social processes of production and consumption (management or "metabolism") and hegemonically defined by social perceptions and interpretations, which, in turn, impose certain limits on these constructions" (Brand & Wissen, 2013: pp. 691–692). Besides the dimensions of biophysical facts and material artefacts, society-nature relations also include a cultural-symbolic dimension. As Ulrich Brand and Markus Wissen exemplify by drawing on the example of cars, cars are material artefacts that have an impact on the natural environment through their production, use, and disposal, but are also related to governmental policies, capitalist forms of competition, and cooperation, and not least to symbolic meanings like status, freedom, progress, etc. (Brand & Wissen, 2013). Society-nature relations are thereby co-constituted by both social and biophysical facts and processes.

Fig. 14.2 illustrates the constitution of society-nature relations through processes of consumption in an abstract and schematic way.

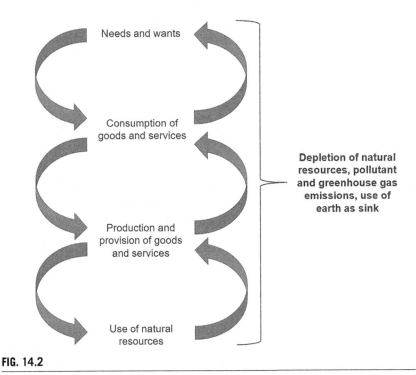

FIG. 14.2

Consumption processes as an element of society-nature relations

Credit: Own illustration.

In Fig. 14.2, the top cycle represents the relationship between culturally shaped needs and wants as basic drivers of consumption and the acquisition and use of goods and services in order to satisfy these needs and wants. While the term "needs" refers to basic essentials for an acceptable standard of living in a given society (e.g., in modern societies: hot water on tap, heating, access to mobility, etc.)[a], "wants" refer to desires for life in excess of needs (e.g., fancy sneakers, designer furniture, the latest technical gadgets, etc.). In modern societies, lifestyles grounded in "taste systems" become the basic generating and structuring principle of wants. While needs are commonly regarded as being finite and satiable, wants are malleable, insatiable, and relative (Slater, 1998). Don Slater exemplifies both the difference between and the close relatedness of needs and wants by drawing on the following example: "If we eat to satisfy the need of hunger, it is argued, we come to an end when the pangs are sated. However, if we consume in order to demonstrate our cosmopolitan status as knowledgeable consumers of the latest culinary styles or choicest ingredients, then the "need" at stake is not hunger but status or prestige, and the latter is always provisional" (Slater, 1998: p. 326).

Before goods and services can be consumed, they must be produced and provided, which is illustrated by the middle cycle in Fig. 14.2. The production and provision of goods and services in turn depend to varying degrees on access to natural resources. In this way, social facts and processes, which mainly shape why and how goods and services are produced, provided, and consumed, are intertwined with biophysical facts and processes.

In sum, these three cycles bring about ecological consequences such as the depletion of natural resources, the generation of pollutant and greenhouse gas emissions, and the use of the earth as a sink. Of course, the scale of the impact on the natural environment depends on the kind of goods and services and on the specific modes of production and provision. The use and potential depletion of critical materials and the socio-ecological consequences of their exploitation are also one facet of society-nature relations.

What becomes apparent here is that consumption is always linked to production and that the ecological consequences of consumption are also grounded in the way the respective goods and services have been produced and provided. This is prominently elaborated in the "system of provision" approach, where a system of provision is understood as "the inclusive chain of activity that attaches consumption to the production that makes it possible" (Fine, 2002: p. 79). This approach stresses that economic and social processes go hand in hand and that consumption and its consequences can only be fully understood when elements located away from the consumption itself, i.e., forms of production, distribution, and retail, are taken into account. This is due to the integral nature of the structures, relationships, and practices of both the demand and the supply side (Fine, 2002). With respect to the critical

---

[a] In its elementary form, the term "needs" denotes basic natural utilities such as food, water, warmth, shelter, etc. As societies develop, needs become greater and less fundamental (Slater, 1998).

materials that are built into goods such as PV systems and EVs during the phase of their production, this perspective reveals that the crucial moment when critical materials are incorporated into the goods is hidden from consumers, but is nevertheless relevant for the socio-ecological impacts of associated acts of consumption.

The "why" of consumption: Sociocultural and sociopsychological functions of goods and services

Consumption acts can be roughly distinguished between (a) inconspicuous consumption, i.e., the "invisible," mostly unconscious use of resources which are not consumed directly (e.g., water use for showering), (b) ordinary consumption, i.e., unspectacular, repetitive, mundane activities (e.g., buying butter or watching TV), and (c) extraordinary consumption, i.e., more or less exceptional activities which require a high degree of conscious personal involvement (e.g., buying a car) (Evans, 2018; Gronow & Warde, 2001). Ordinary and extraordinary consumption can be further differentiated between low-involvement and high-involvement consumption, whereby high-involvement activities or products are especially laden with personal and social meaning and are therefore of higher significance for defining self-identity (Belk, 1995). However, the degree of involvement depends on both situative and individual factors (e.g., personality traits, income, past experiences, etc.) and may also vary across the different stages of the consumption process.

While inconspicuous and ordinary consumption are mainly, but not exclusively, associated with the everyday use of goods, extraordinary consumption applies mainly to the acquisition of goods. The acquisition of PV systems and EVs thus resides in the realm of extraordinary consumption, while the electricity consumption associated with the use of an EV can be regarded as inconspicuous consumption and its everyday use as ordinary consumption. As this example shows, different modes of consumption can apply to the same good depending on the respective stage in the consumption process.

As mentioned above, needs and wants are basic motivators of consumption. In order to gain a deeper sociological understanding of the human motivation to consume different goods and services, it is helpful to refer to the functions these goods and services (promise to) fulfill (Jackson, 2006). These functions are always of a social nature: the functions that goods and services are supposed to fulfill are socially constructed and thus not inherent to the respective goods and services. Furthermore, consumption activities are embedded in sociopolitical constellations. For example, buying and using a PV system involves profiting from feed-in tariffs, contributing to the success of the renewable energy technology industry, changing the electricity mix in the grid, etc. Thus, being an owner of a PV system positions the individual within a certain sociopolitical constellation (i.e., discourses, policies, power relations in the context of energy transitions).

In the sociologically oriented literature on consumption, the functions of consumption are classified in different ways (e.g., Campbell, 1995; Reisch, 2002; Richins, 1994). The following presents and describes the five most common functions:

- **Utilitarian function**: First of all, most goods and services fulfill basic utilitarian purposes for the consuming individual. For instance, cars are a means of covering distances without much physical effort, and food is used for stilling hunger. The utilitarian function thus refers to the practical use value of goods and services.
- **Hedonistic function**: Goods and services are also a means of experiencing enjoyment. Individuals derive pleasure from the acquisition and use of goods and services. The consumption of experiences, such as the attendance of concerts, provide particularly illustrative examples of the hedonistic function. Also the mere act of shopping is a source of enjoyment for many people.
- **Positional function**: As most prominently elaborated by Pierre Bourdieu (Bourdieu, 1984) and Thorstein Veblen (Veblen, 2007[1899]), goods and services are also used for displaying one's social position in society and thus distinguishing the respective owner or user from others in terms of his or her social status. An obvious example is the use of cars as status symbols that make one's wealth and success visible for other members of society. The display of status, however, can also take more subtle forms: visiting the opera can be used to symbolically distinguish oneself from the less educated petty bourgeois.
- **Integrative function**: Besides being used to make status distinctions between the social strata, goods and services also fulfill an integrative function. They are the symbols of group belonging and therefore manifestations of social order (Douglas & Isherwood, 1996[1979]). For instance, the possession of a vintage car can be the "entrance card" to a specific group of car enthusiasts and thus identifies the owner as part of this actual or imagined group. Further example, the extensive consumption of meat can be regarded as a sign of masculinity in some social groups, thus manifesting and reproducing a gender-specific social order.
- **Expressive function**: The acquisition and use of certain goods and services also provide a means for individuals to express their identity and personal values. For example, the possession of a PV system can be part of the owner's way of expressing his or her "green" identity. As Zygmunt Bauman has elaborated (Bauman, 2001), there is a mutual fit between consumer culture and the specific conditions of modernity under which individuals are no longer endowed with a stable (and largely immutable) identity by virtue of being born into a specific social position. In modernity, individuals are instead put in the position of having to constantly construct and stabilize their identity throughout the course of their lives. Consumer goods such as clothes or furniture can therefore be a means of identity construction and expression.

Of course, most goods and services can fulfill different functions at once. Furthermore, as the positional, integrative, and expressive functions illustrate, goods and services carry symbolic meaning and are thus used as communicational tools conveying messages about the owner's or user's identity, values, social status, preferred social order, group belonging, and so on. In order to effectively function as carriers of symbolic meaning, the following two requirements must be met: visibility and significance. In other words, the (use of) specific goods and services must be visible to others to some extent and the symbolic meaning must be collectively shared, so that others are able to decode and thus understand the symbolic value of the respective good or service (Wiswede, 2000).

In the context of ongoing mobility and energy transitions, consumers are regarded and mobilized by different stakeholders (policy-makers, business actors, NGOs, etc.) as central actors in the "greening" of energy and mobility systems (Evans, Welch, & Swaffield, 2017). Furthermore, goods such as PV systems or EVs and their diffusion among wider parts of society are discursively presented as imperatives for the success of sustainability transitions (Wentland, 2016). Accordingly, the framing of such goods as crucial elements in sustainability transitions becomes anchored in consumers' images of such transitions (Böhm, Doran, & Pfister, 2018). These goods thus become carriers of social meaning: they symbolize the potential for a more sustainable future. As we will see in the following section, environmental protection has become a significant motive for the acquisition of such goods.

The case of electric vehicles and photovoltaic systems

In the context of renewable energy transitions, PV systems and EVs are the most obvious examples of technologies that are acquired by private households and contain critical materials (Hofmann, Hofmann, Hagelüken, & Hool, 2018). Lithium, cobalt, nickel, and graphite are used in battery systems, while electric motors feature dysprosium, neodymium, and praseodymium, and thin film photovoltaic modules are made using silver, gallium, indium, and tellurium (Hofmann et al., 2018; Wäger, Lang, Wittmer, Bleischwitz, & Hagelüken, 2012). Although many of the critical materials used in renewable energy technologies are not scarce (Overland, 2019), sustainability issues arise from the way they are currently mined (see Phadke, McLellan, and Wang et al. within this book). Common consequences and issues associated with the mining of critical materials are loss of farmland, air pollution, contamination of water supplies, highly energy-intensive and water-intensive extraction processes, as well as serious health risks for mining workers (Brunnengräber & Haas, 2018; Faria et al., 2013). Thus, while EVs and PV systems are supposed to support the "greening" of lifestyles in Western countries, the costs of producing the necessary raw materials are externalized to other countries (Brand & Wissen, 2018a).

The socio-ecological impacts associated with critical materials embodied in EVs and PV systems raise crucial questions concerning the demand side. Are consumers aware of these issues, and if so, does it actually make a difference? To begin answering these questions, I will briefly review the state of research on the main factors driving the adoption of EVs and PV systems (for comprehensive and detailed overviews please see, for example, Rezvani, Jansson, & Bodin, 2015 for EVs and Wolske, Stern, & Dietz, 2017 for PV systems).

The adopters of both EVs and PV systems roughly match the general sociodemographic characteristics of early adopters in Everett Rogers' diffusion of innovations theory (Rogers, 2003: p. 288): they are well educated and financially well situated. Moreover, the "demonstration of innovativeness" has been identified as an important motive for both EV and PV system adopters. It can thus be assumed that the current research into these driving factors is mainly looking at "early adopters," in relation to both EVs and PV systems. As these innovations become popular among different consumer brackets, other drivers are likely to gain importance. This overview must therefore be regarded as a snapshot in the diffusion trajectory of these innovations.

The overview presented in Table 14.1 also makes it evident that the acquisition of EVs and PV systems particularly addresses the utilitarian, positional, and expressive functions of consumption. The adopters of EVs and PV systems expect financial gains, positive environmental effects, and also—at least in the case of EVs—gains in status. As a motive, "environmental protection" can thereby refer to both the utilitarian and the expressive functions of these goods. EVs and PV systems can be acquired with the aim of contributing to environmental protection (utilitarian function) and expressing one's "green" identity (expressive function). As research has shown, environmental motivations are neither necessary nor sufficient for the adoption of EVs and PV systems (Axsen, TyreeHageman, & Lentz, 2012; Schelly, 2014; Sonnberger, 2015). Thus, adopters may just as well be exclusively motivated by the expected

Table 14.1 Drivers of the adoption of EVs and PV systems.

EV adoption (Axsen & Sovacool, 2019; Boudet, 2019; Li, Long, Chen, & Geng, 2017; Rezvani et al., 2015)	PV system adoption (Boudet, 2019; Schelly, 2014; Sonnberger, 2015: pp. 18–21; Wolske et al., 2017)
• Well-educated, relatively high-income level, engaged in technical professions • Positive effect of environmental awareness • Positive effect of previous (positive) experiences with EVs • Positive effect of financial incentives (tax incentives, subsidies) • Positive effect of expected fuel cost savings • Negative effect of relatively high purchase costs (compared to diesel or petrol cars) • Negative effect of perceived unavailability of charging infrastructure • Negative effect of range anxiety • Demonstration of innovativeness as a motive • Expected status gains as a motive	• Middle-aged, male, well-educated, relatively high-income level, engaged in technical professions • Positive effect of environmental awareness • Positive effect of trust in installers • Positive effect of adopters in reference groups (neighbors, friends or relatives) • Positive effect of financial incentives (expected financial returns from feed-in tariffs, subsidies) • Demonstration of innovativeness as a motive • Mixed evidence concerning expected status gains as a motive

financial gains. Furthermore, environmental concerns may also constitute a barrier for the adoption of EVs and PV systems if people do not consider the technologies to be environmentally beneficial. This can be based on life-cycle considerations, when consumers assume that pollution is created in the process of production or disposal (Axsen et al., 2012; Sonnberger, 2015). However, there is no empirical evidence that this is explicitly the case with regard to critical materials.

Not surprisingly, demonstrating one's status is an important motive for the adoption of EVs, since cars in general are one of the most powerful status symbols in modern societies. As high involvement goods, cars are also closely linked to expressions of identity that result in milieu-specific preferences for certain automobile brands and types. Since PV systems lack this long-standing status as symbolic signifiers and the PV product range is far less diverse than that of cars, their function as status symbols may be less pronounced.

To sum up, as research has shown, symbolic aspects such as self-expression and demonstrations of social status matter in both cases. Both goods are mainly perceived as being environmentally beneficial. However, some consumers also seem to shun the adoption of EVs and PV systems for environmental reasons (based on life-cycle considerations), although they are most likely a small minority. There is no empirical evidence that embodied critical materials are viewed as an issue among larger segments of potential adopters. Environmental protection, financial incentives and expected financial gains, demonstrations of status, and expressions of a "green" and/or innovative identity are the dominant motives.

Discussion and conclusion

Sustainability issues related to critical materials embodied in EVs and PV systems are grounded in the moment of acquisition, the processes of production before the acquisition, and in the moment of disposal and what comes after it. Since embodied critical materials are invisible to consumers and a public debate about this issue is in its infancy (if it exists at all), some scholars argue for making embodied materials more salient to consumers and raising public awareness in general (Liu, Wang, & Su, 2016). Product labeling is therefore regarded by some as a promising approach, as it helps put potential consumers in a position to make informed purchasing decisions. Underlying this is the hope that consumers will use their purchasing power to force industry players to consider sustainability aspects throughout the whole supply chain and provide more sustainable products, thereby pointing the respective society-nature relations in a more sustainable direction. With regard to the disposal of PV systems and EVs, recycling of embodied critical materials is also an important issue. Critical materials are not usually simply consumed in a practical sense, since they are still materially present after they have been used. Thus, they can be—at least partly—recycled. However, the efficiency of recycling activities depends on the way the respective products are disposed of (see Weil et al., and Goddin, Chapter 12 within this book). Furthermore, if products contain only small amounts of critical materials, recycling is neither economically nor technically feasible (Bulach et al., 2018). Besides labeling that addresses the moment of acquisition, awareness raising about the importance of recycling at the moment of disposal is another potential strategy for enhancing product sustainability throughout the whole life cycle (Lapko, Trianni, Nuur, & Masi, 2019). Given that consumers seem to be mostly unaware of the critical materials that EVs and PV systems are based on, providing information sounds like an obvious and reasonable approach. However, information provision strategies to steer behavior in more sustainable directions are rooted in the so-called "information deficit" model, which has been widely critiqued for its underlying conception of consumers as rational actors who behave more sustainably as soon as they have the relevant information about the negative impacts of their behavior (Hinton & Goodman, 2010). Studies have shown that additional and new information is always processed on the basis of existing attitudes and thus in a biased way (Renn, 2008). Furthermore, due to the complexity of sustainability issues, "expert dilemma" exist in many cases making it difficult to decide which information is correct. This often results in the politicization of information, as consumers are left confused and irritated.

From a critical perspective, another crucial objection against addressing consumers with information strategies and awareness raising campaigns that promote sustainable consumption is that such techniques individualize responsibility for socio-ecological issues and necessary social change (Evans et al., 2017; Shove, 2010). This reinforces the idea that citizens could and should bring about socio-ecological change on their own, thereby detracting from the responsibility of industry players and policy-makers to establish more sustainable production systems and policy frameworks. As Emma Hinton and Michael Goodman put it: "merely providing individuals with information relating to SC [=sustainable consumption] fails to tackle the roots of society's lock-in to high-consumption lifestyles in terms of its economic, technological, and cultural groundings" (Hinton & Goodman, 2010: p. 250). With regard to consumption patterns and critical materials, other scholars also emphasize that the transformation of respective society-nature relations requires far-reaching changes ranging from technological optimization and policy regulations to the adaptation of lifestyles (Wäger et al., 2012). Accordingly, Patrick Wäger and his colleagues stress the importance of sufficiency strategies (Wäger et al., 2012). This conclusion is, for example, supported by a recent study which estimated the energy costs of a transition from the current transportation system to a 100% renewable transportation system worldwide (García-Olivares, Solé, & Osychenko, 2018). The authors show that this would be feasible. However, due to various factors such as the limited availability of critical materials, the feasibility would depend on accompanying sufficiency measures in the field of private car ownership and use. Thus, they recommend a limited use of EVs, which would equate to a relatively small fleet (García-Olivares et al., 2018). Accordingly, scholars such as Ulrich Brand and Markus Wissen stress that the selective ecological modernization of dominant patterns of consumption without sufficiency will not help to solve socio-ecological issues on a global scale, as long as the negative side effects of the "greening" of (Western) lifestyles—such as the socio-ecological consequences of the extraction of critical materials—are externalized to other regions in the world (Brand & Wissen, 2018b).

The adoption and diffusion of EVs and PV systems is also politically desired in many countries as a means to mitigate climate change and is thus supported by policies that provide financial incentives. Furthermore, although critical voices exist, EVs and PV systems are mainly framed as environmentally friendly in public debates, which is further reinforced by respective marketing strategies. As shown earlier, the adoption of EVs and PV systems is driven by a multitude of factors, including symbolic meanings that motivate people to adopt the technologies in order to express themselves or demonstrate their social status. Even if consumers were aware of the issue of critical materials embodied in these technologies, it is questionable whether this would have an impact on their adoption decisions. Of course, it is conceivable that the emergence of a public debate about critical materials could trigger a reframing of EVs and PV systems. Such reframing would enable people to put pressure on producers to ensure the sustainability of their products throughout the whole supply chain and life cycle. However, this would also dissuade people from buying EVs and PV systems due to environmental and social concerns, which in turn would have negative side effects with regard to the reduction in CO_2 emissions in the electricity and transport sectors.

In sum, the case of critical materials embodied in those technologies is an excellent reminder that "sustainable consumption is characterized by a focus on consuming less rather than consuming differently" (Evans et al., 2017: p. 1409) and that the socio-ecological impacts of consumption are always inevitably linked to the way goods and services are produced and supplied. Together, these elements constitute a specific configuration of society-nature relations that eventually varies in its degree of sustainability.

References

Axsen, J., & Sovacool, B. K. (2019). The roles of users in electric, shared and automated mobility transitions. *Transportation Research Part D: Transport and Environment, 71*, 1–21. https://doi.org/10.1016/j.trd.2019.02.012.

Axsen, J., TyreeHageman, J., & Lentz, A. (2012). Lifestyle practices and pro-environmental technology. *Ecological Economics, 82*, 64–74. https://doi.org/10.1016/j.ecolecon.2012.07.013.

Bauman, Z. (2001). Consuming life. *Journal of Consumer Culture, 1*(1), 9–29.

Becker, E., Hummel, D., & Jahn, T. (2011). Gesellschaftliche Naturverhältnisse als Rahmenkonzept. In M. Groß (Ed.), *Handbuch Umweltsoziologie* (pp. 75–96). Wiesbaden: VS Verlag für Sozialwissenschaften.

Belk, R. W. (1995). Studies in the new consumer behavior. In D. Miller (Ed.), *Acknowledging consumption. A review of new studies* (pp. 58–95). London, New York: Routledge.

Böhm, G., Doran, R., & Pfister, H.-R. (2018). Laypeople's affective images of energy transition pathways. *Frontiers in Psychology, 9*, 1–15. https://doi.org/10.3389/fpsyg.2018.01904.

Boudet, H. S. (2019). Public perceptions of and responses to new energy technologies. *Nature Energy, 29*, 135. https://doi.org/10.1038/s41560-019-0399-x.

Bourdieu, P. (1984). *Distinction: A social critique of the judgement of taste*. Cambridge: Harvard University Press.

Brand, U., & Wissen, M. (2013). Crisis and continuity of capitalist society-nature relationships: The imperial mode of living and the limits to environmental governance. *Review of International Political Economy, 20*(4), 687–711. https://doi.org/10.1080/09692290.2012.691077.

Brand, U., & Wissen, M. (2018a). What kind of great transformation? The imperial mode of living as a major obstacle to sustainability politics. *GAIA—Ecological Perspectives for Science and Society, 27*(3), 287–292. https://doi.org/10.14512/gaia.27.3.8.

Brand, U., & Wissen, M. (2018b). *The limits to capitalist nature. Theorizing and overcoming the imperial mode of living*. London: Rowman & Littlefield International.

Brunnengräber, A., & Haas, T. (2018). Vom Regen in die Traufe: die sozial-ökologischen Schattenseiten der E-Mobilität. *GAIA—Ecological Perspectives for Science and Society, 27*(3), 273–276. https://doi.org/10.14512/gaia.27.3.4.

Bulach, W., Schüler, D., Sellin, G., Elwert, T., Schmid, D., Goldmann, D., et al. (2018). Electric vehicle recycling 2020: Key component power electronics. *Waste Management & Research, 36*(4), 311–320. https://doi.org/10.1177/0734242X18759191.

Campbell, C. (1995). The sociology of consumption. In D. Miller (Ed.), *Acknowledging consumption. A review of new studies* (pp. 96–126). London, New York: Routledge.

Defila, R., Di Giulio, A., & Kaufmann-Hayoz, R. (2012). Introduction. In R. Defila, A. Di Giulio, & R. Kaufmann-Hayoz (Eds.), *The nature of sustainable consumption and how to achieve it. Results from the focal topic "From knowledge to action—New paths towards sustainable consumption"* (pp. 11–20). Munich: oekom.

Douglas, M. T., & Isherwood, B. (1996 [1979]). *The world of goods: Towards an anthropology of consumption*. London: Routledge.

Ellsworth-Krebs, K., & Reid, L. (2016). Conceptualising energy prosumption: Exploring energy production, consumption and microgeneration in Scotland, UK. *Environment and Planning A, 48*(10), 1988–2005. https://doi.org/10.1177/0308518X16649182.

Evans, D. M. (2018). What is consumption, where has it been going, and does it still matter? *The Sociological Review, 65*, 1–19. https://doi.org/10.1177/0038026118764028.

Evans, D., Welch, D., & Swaffield, J. (2017). Constructing and mobilizing 'the consumer': Responsibility, consumption and the politics of sustainability. *Environment and Planning A, 49*(6), 1396–1412. https://doi.org/10.1177/0308518X17694030.

Faria, R., Marques, P., Moura, P., Freire, F., Delgado, J., & de Almeida, A. T. (2013). Impact of the electricity mix and use profile in the life-cycle assessment of electric vehicles. *Renewable and Sustainable Energy Reviews, 24*, 271–287. https://doi.org/10.1016/j.rser.2013.03.063.

Fine, B. (2002). *The world of consumption: The material and cultural revisited* (2nd ed.). London: Routledge.

Fischer, D., Michelsen, G., Blättel-Mink, B., & Di Giulio, A. (2012). Sustainable consumption: how to evaluate sustainability in consumption acts. In R. Defila, A. Di Giulio, & R. Kaufmann-Hayoz (Eds.), *The nature of sustainable consumption and how to achieve it. Results from the focal topic "From knowledge to action—New paths towards sustainable consumption"* (pp. 67–80). Munich: oekom.

García-Olivares, A., Solé, J., & Osychenko, O. (2018). Transportation in a 100% renewable energy system. *Energy Conversion and Management, 158*, 266–285. https://doi.org/10.1016/j.enconman.2017.12.053.

Gronow, J., & Warde, A. (2001). Introduction. In J. Gronow & A. Warde (Eds.), *Ordinary consumption* (pp. 1–8). London: Routledge.

Hinton, E. D., & Goodman, M. K. (2010). Sustainable consumption: Developments, considerations and new directions. In M. R. Redclift & G. Woodgate (Eds.), *The international handbook of environmental sociology* (2nd ed., pp. 245–261). Cheltenham: Edward Elgar.

Hofmann, M., Hofmann, H., Hagelüken, C., & Hool, A. (2018). Critical raw materials: A perspective from the materials science community. *Sustainable Materials and Technologies, 17*, 1–10. https://doi.org/10.1016/j.susmat.2018.e00074.

International Energy Agency. (2018). *Global EV outlook 2018: Towards cross-modal electrification*. Paris: International Energy Agency.

Jackson, T. (2006). Consuming paradise? Towards a social and cultural psychology of sustainable consumption. In T. Jackson (Ed.), *The Earthscan reader in sustainable consumption* (pp. 367–395). London, Sterling: Earthscan.

Kollmuss, A., & Agyeman, J. (2002). Mind the gap: Why do people act environmentally and what are the barriers to pro-environmental behavior? *Environmental Education Research*, *8*(3), 239–260. https://doi.org/10.1080/13504620220145401.

Lapko, Y., Trianni, A., Nuur, C., & Masi, D. (2019). In pursuit of closed-loop supply chains for critical materials: An exploratory study in the green energy sector. *Journal of Industrial Ecology*, *23*(1), 182–196. https://doi.org/10.1111/jiec.12741.

Li, W., Long, R., Chen, H., & Geng, J. (2017). A review of factors influencing consumer intentions to adopt battery electric vehicles. *Renewable and Sustainable Energy Reviews*, *78*, 318–328. https://doi.org/10.1016/j.rser.2017.04.076.

Liu, T., Wang, Q., & Su, B. (2016). A review of carbon labeling: Standards, implementation, and impact. *Renewable and Sustainable Energy Reviews*, *53*, 68–79. https://doi.org/10.1016/j.rser.2015.08.050.

Moser, S., & Kleinhückelkotten, S. (2018). Good intents, but low impacts: Diverging importance of motivational and socioeconomic determinants explaining pro-environmental behavior, energy use, and carbon footprint. *Environment and Behavior*, *50*(6), 626–656. https://doi.org/10.1177/0013916517710685.

Overland, I. (2019). The geopolitics of renewable energy: Debunking four emerging myths. *Energy Research & Social Science*, *49*, 36–40. https://doi.org/10.1016/j.erss.2018.10.018.

Partnership on Sustainable Low Carbon Transport. (2019). *E-mobility trends and targets: As of March 15, 2019 (constantly updated)*. http://slocat.net/sites/default/files/e-mobility_overview.pdf. (Accessed 7 May, 2019).

Raman, S. (2013). Fossilizing renewable energies. *Science as Culture*, *22*(2), 172–180. https://doi.org/10.1080/09505431.2013.786998.

Reisch, L. A. (2002). Symbols for Sale: Funktionen des symbolischen Konsums. In C. Deutschmann (Ed.), *Die gesellschaftliche Macht des Geldes* (pp. 226–248). Wiesbaden: Westdeutscher Verlag.

Renewable Energy Policy Network. (2018). *Renewables 2018: Global status report*. Paris: REN21 Secretariat.

Renn, O. (2008). *Risk governance: Coping with uncertainty in a complex world*. London: Earthscan.

Rezvani, Z., Jansson, J., & Bodin, J. (2015). Advances in consumer electric vehicle adoption research: A review and research agenda. *Transportation Research Part D: Transport and Environment*, *34*, 122–136. https://doi.org/10.1016/j.trd.2014.10.010.

Richins, M. L. (1994). Valuing things: The public and private meanings of possessions. *Journal of Consumer Research*, *21*(3), 504–521.

Rogers, E. M. (2003). *Diffusion of innovations* (5th ed.). New York, London, Toronto, Sydney: The Free Press.

Sanne, C. (2002). Willing consumers—or locked-in? Policies for a sustainable consumption. *Ecological Economics*, *42*, 273–287.

Schelly, C. (2014). Residential solar electricity adoption: What motivates, and what matters? A case study of early adopters. *Energy Research & Social Science*, *2*, 183–191. https://doi.org/10.1016/j.erss.2014.01.001.

Schneider, N. F. (2000). Konsum und Gesellschaft. In D. Rosenkranz & N. F. Schneider (Eds.), *Konsum. Soziologische, ökonomische und psychologische Perspektiven* (pp. 9–22). Opladen: Leske + Budrich.

Shove, E. (2010). Beyond the ABC: Climate change policy and theories of social change. *Environment and Planning, 42*(6), 1273–1285. https://doi.org/10.1068/a42282.

Slater, D. (1998). Needs/wants. In C. Jenks (Ed.), *Core sociological dichotomies* (pp. 315–328). London: Sage.

Sonnberger, M. (2015). *Der Erwerb von Photovoltaikanlagen in Privathaushalten: Eine empirische Untersuchung der Handlungsmotive, Treiber und Hemmnisse.* Wiesbaden: Springer VS.

Stern, P. C. (2000). Toward a coherent theory of environmentally significant behavior. *Journal of Social Issues, 56*(3), 407–424.

Veblen, T. (2007 [1899]). *The theory of the leisure class.* Oxford: Oxford University Press.

Wäger, P. A., Lang, D. J., Wittmer, D., Bleischwitz, R., & Hagelüken, C. (2012). Towards a more sustainable use of scarce metals: A review of intervention options along the metals life cycle. *GAIA—Ecological Perspectives for Science and Society, 21*(4), 300–309. https://doi.org/10.14512/gaia.21.4.15.

Warde, A. (2005). Consumption and theories of practice. *Journal of Consumer Culture, 5,* 131–153.

Wentland, A. (2016). Imagining and enacting the future of the German energy transition: Electric vehicles as grid infrastructure. *Innovation: The European Journal of Social Science Research, 29*(3), 285–302. https://doi.org/10.1080/13511610.2016.1159946.

Wiswede, G. (2000). Konsumsoziologie—Eine vergessene Disziplin. In D. Rosenkranz & N. F. Schneider (Eds.), *Konsum. Soziologische, ökonomische und psychologische Perspektiven* (pp. 23–94). Opladen: Leske + Budrich.

Wolske, K. S., Stern, P. C., & Dietz, T. (2017). Explaining interest in adopting residential solar photovoltaic systems in the United States: Toward an integration of behavioral theories. *Energy Research & Social Science, 25,* 134–151. https://doi.org/10.1016/j.erss.2016.12.023.

Renewable energy and critical minerals: A field worthy of interdisciplinary research

Alexandra Pehlken[a] and Alena Bleicher[b]

[a]*OFFIS—Institute for Information Technology, Oldenburg, Germany*
[b]*Helmholtz Centre for Environmental Research—UFZ, Leipzig, Germany*

This book started with the observation that the relationship between renewable energy systems and their material basis has so far been somewhat neglected in scientific research, or at least, only addressed from rather narrow disciplinary perspectives. The book cover reflects the relationship between the so-called green renewable energies and the often environmentally damaging mining operations on which those technologies rely. In order to paint a broader picture, we invited authors with a variety of disciplinary backgrounds to shed light on this relationship using examples from different regions around the world.

The chapters within this book reveal manifold issues that are relevant to the material basis of renewable energies. They also make clear that advanced research results are available on some aspects of the material basis of renewable energies. However, these results usually have a strong disciplinary focus. An interdisciplinary perspective on the issue, beyond teaching purposes, is currently lacking (SusCritMat, 2020). The chapters raise topics that would be worth researching from an interdisciplinary perspective. We have selected some of these overarching issues and discuss their potential for interdisciplinary research.

Several of the authors explain that it is difficult to anticipate the future development of material needs. This is due to many unknown variables, such as the pathways and future development of existing or emerging energy technologies, uncertain geopolitical constellations, as well as national resource and energy policies and their dynamics that directly impact the availability of resources. Also, the output produced by assessment methods varies according to the assumptions, default settings, and indicators that are applied. The lack of available information makes it difficult to assess the material requirements and long-term sustainability of renewable energy technologies (see Penaherrera and Pehlken, Chapter 8). Against this background, an overarching conclusion can be drawn: there is ongoing uncertainty about the amount and type of materials that will be needed for renewable energy technologies in the future, and the locations from which they will be sourced. This leaves wiggle room and

The Material Basis of Energy Transitions. https://doi.org/10.1016/B978-0-12-819534-5.00015-5

thus, the authors in this book come to different conclusions. For example, Wang et al. (Chapter 3) argue that the physical scarcity of minerals will not be a major concern for energy technologies in the future, while Weil et al. (Chapter 5) emphasize that the limited availability of materials is a major issue in the context of energy storage using battery technologies.

In spite of such divergent conclusions, the authors are united by their overall goal, which is to apply the criteria of sustainable development (social, ecological, economic) to the production and use of materials needed for renewable energy technologies. Sustainability criteria are relevant for technological developments such as the use of renewable energies in mining, the development of technologies and concepts for recycling, and the search for substitutes for materials that are either environmentally harmful or will become scarce and expensive in the near future. To date, economic importance and supply risk are still the most common sets of parameters used in criticality assessments. Koch (Chapter 9) states that more discussion is needed about the moral and ethical aspects related to the resource consumption of renewable energies. He argues that we have a moral and ethical obligation to ask more questions about sustainable mining and production, in order to establish sustainable energy systems. A renewable energy system may turn out to be the most sustainable in one particular country due to its geographical (e.g., long insolation), economic (e.g., available expertise and materials), or geophysical structures (e.g., location near to tectonic plate boundaries that make geothermal energy accessible or nearby availability of needed materials for construction). In addition, social (e.g., openness of industries and the public toward renewable energy technology) or political (e.g., subsidies or regulatory frameworks) structures in different countries influence the shape of renewable energy systems—centralized or decentralized production and supply system, dominance of single or multiple energy sources, etc. For example, a country such as China, which is rich in rare-earth elements, has direct access to raw materials and can more easily control the market conditions and even impact the environmental conditions of raw material production (e.g., for photovoltaic cells). In contrast, European countries have to import most of the raw materials required for solar energy technology. The longer supply chain means it is more difficult to ensure the sustainability of the materials. This leads to an odd situation in which a given technology is sustainable in one country, yet the same system may be completely unsustainable in another country (see Zepf, Chapter 4). The sustainability of energy systems should be analyzed from a life cycle perspective. Life cycle assessment (LCA) methods are relevant in this regard; they are used to assess the whole product chain. LCAs are currently used to evaluate mineral criticality as an indicator of the potential environmental risk associated with certain metal supply chains. However, several related problems have been identified by the authors in this book. A major critique is voiced by McLellan (Chapter 7), who argues that the dominant perspective on environmental aspects is a global perspective. Less attention is devoted to the local environmental impacts of minerals and their supply chains—from the mine to the manufacturing of the product (and potentially through to recycling or disposal)—but such impacts are highly relevant and should, therefore, be taken into consideration to a far greater extent. This could be done using frameworks, such as the Environmental Impact Assessment (EIA), which are embedded in common regulatory processes and often made available for public comment. Expanding the analysis of the impacts of material consumption requires an evaluation of the sociopolitical, economic, and environmental dimensions. Relevant indicators have to be developed and added to the existing LCA and social LCA approaches. In order to achieve this goal, expertise from different fields of the natural sciences, engineering, and social sciences is required to identify the relevant social and environmental processes and developments that should be monitored by indicator systems. Examples here include climate-relevant emissions, resource consumption, supply chain, involvement of stakeholders and consumers, subsidies, and human rights, which are often integrated into global reporting guidelines for sustainability reporting.

The authors suggest various strategies that could be used to build sustainable energy systems, such as better sector coupling (e.g., electricity combined with heat and mobility), the use of other flexibility options (e.g., load management and improved predictions of energy production), and the reduction of total electricity consumption (Weil et al., Chapter 5). Their discussions focus primarily on technological solutions. An interdisciplinary perspective could reveal the interplay between technical solutions, natural conditions, and social and political dynamics in nation states and economies. This perspective has the potential to open up all kinds of new opportunities, such as smart cities with a sharing economy that is combined with sector coupling in the form of a private energy cooperative.

Today, there is a clear shift away from fossil fuels and toward new types of energy systems worldwide, as demand for energy continues to grow in all countries. This energy transition is linked to a variety of conflicts related to the raw materials that are required for green technologies. However, conflicts and questions of environmental justice related to mining operations are often invisible in debates on the operation of renewable energy technologies. Several chapters within this book aim to make the readers aware of these issues (Prause, Chapter 10, Koch, Chapter 9, Phadke, Chapter 2, McLellan, Chapter 7, Gilbert, Chapter 6). Thus, investment in renewable energies should consider the fairness of the supply chain. As Prause highlights, there needs to be more research into the dynamics of resource conflicts that involve actors in resource-producing countries (trade unions, miners, and local governments), and in countries of the Global North (industries such as automotive, social movements contesting the expansion or implementation of industrial mining in the Global South).

In this context, the interdependency between climate, energy, resource, and waste policies also needs to be studied, so that we can better understand their impact on technology development pathways, related resource needs, and the dynamics of resource conflicts. While social scientists can contribute more detailed knowledge about structures and dynamics between local, national, and international decision-making (e.g., reinforcing conflict dynamics), an engineering and natural scientific approach can indicate consequences related to the amount of materials needed, mining and recycling activities, and even the limits of recycling. An understanding of these interrelationships can then be used to develop and refine management strategies and political measures. Such research is important for gaining a better understanding of how resource conflicts will be impacted by greater demand for minerals in the renewable energy technology sector and related policies. It will also facilitate the development of governance instruments that make supply chains more sustainable.

New governance instruments such as certification schemes are required to ensure transparency along trading and product chains, incorporate aspects relevant in mining areas, and solve resource-related conflicts. Governance instruments have so far focused on product chains, and address mineral producers and technology producers. In this regard, the combination of the perspectives taken by Gilbert, Phadke, and Franken et al. in this book seems promising (Chapters 6, 2, and 11). They question the extent to which the values and imaginaries that underlie the various certification initiatives actually enable the incorporation of local perspectives, such as the environmental justice concerns of local communities, or the working conditions in the artisanal mining sector. Based on such an understanding, existing certification schemes—as well as concepts such as criticality, sustainable mining, and responsible mining—can be redefined, extended, and practiced within a collaborative process. This approach would make it possible to improve mining practices and to harmonize local, national, and global development goals.

The circular economy is an approach that can benefit from interdisciplinary research methods and research conducted from a socio-technical perspective (Graedel, Reck, Ciacci, & Passarini, 2019). In the context of renewable energy systems, a circular economy refers to the implementation of recycling strategies for renewable energy technologies, the identification and development of substitutes (Goddin, Chapter 13), the enhancement of material efficiency, as well as the reduction of demand for specific technologies (e.g., stationary batteries) (Weil et al., Chapter 5). Besides the technical aspects and solutions such as recycling technologies, extending a product's service life, improving product durability, or developing new materials (Goddin, Chapter 13), sociocultural structures, and dynamics are also relevant to circular economies and need to be investigated. Recycling strategies related to urban mines (here: already constructed and operating renewable energy technologies) are considered to be highly efficient, because all the necessary information about the materials they contain is available today and in the future.

Recycling technologies in a circular economy require individual consumers to develop (new) waste-separating practices, as well as the development of appropriate waste collection systems. On the consumer side, practices of sorting waste are relevant in the context of renewable energy technologies, as owners of decentralized facilities such as private solar PV systems are responsible for the correct dismantling and delivery of their systems to recycling facilities. [For example, in Europe, end-of-life photovoltaic panels should be taken to facilitate that process Waste Electrical & Electronic Equipment (WEEE).] Furthermore, new practices such as sharing objects and products (e.g., battery sharing) are becoming important in this regard (e.g., Osburg, Strack, & Toporowski, 2016). At the same time, it is important to be aware of the impact (or lack of impact) that consumers' choices and practices can have on the material basis of renewable energy technologies. As Sonnberger (Chapter 14) reveals, the production phase is the most important phase in this context. It is also necessary to take into account the fact that, in future, the number of smaller energy producers like photovoltaic panels on private roofs and neighborhood-scale energy systems will increase and potentially become the dominant structure.

On the producer side, it is important to know why and under which conditions the use of new or recycled products is acceptable (e.g., Bleicher, 2020; Levidow & Raman, 2019; Wittstock, Pehlken, & Wark, 2016). Product design is a social process in which the interests, value systems, and practices of different social actors (e.g., engineers, designers, and economists) are negotiated and inscribed in technologies and products (e.g., the durability of products). James Goddin (Chapters 12 and 13) and other authors (e.g., Bahrudin, Aurisicchio, & Baxter, 2017; Piselli, Baxter, Simonato, Del Curto, & Aurisicchio, 2018) make it clear that implementing a circular economy requires changes in the mindset and practices of designers and engineers. For example, a product's end-of-life phase needs to be considered during the design process (Kalverkamp, Pehlken, & Wuest, 2017; Reuter, van Schaik, Gutzmer, Bartie, & Abadías-Llamas, 2019). Finally, as Phadke highlights (Chapter 2), it is relevant to ask what kind of planning, business incentives, and infrastructures are necessary to establish a circular economy.

Thus, the material basis of renewable energy technologies is a field worthy of further exploration from an interdisciplinary research perspective. An interdisciplinary approach will create a knowledge base that will facilitate sound political decision-making, and make a long-term contribution to the mitigation of climate change.

References

Bahrudin, F. I., Aurisicchio, M., & Baxter, W. L. (2017). *Sustainable materials in design projects*. In *Conference paper presented at the 2017 international conference organized by the Design Research Society Special Interest Group on Experiential Knowledge (EKSIG)*.

Bleicher, A. (2020). Why are recycled waste materials used reluctantly? Enriching research in recycling with social scientific perspectives. *Resources, Conservation and Recycling, 152*, 104543. https://doi.org/10.1016/j.resconrec.2019.104543.

Graedel, T. E., Reck, B. K., Ciacci, L., & Passarini, F. (2019). On the spatial dimension of the circular economy. *Resources, 8*(1), 32. https://doi.org/10.3390/resources8010032.

Kalverkamp, M., Pehlken, A., & Wuest, T. (2017). Cascade use and the management of product lifecycles. *Sustainability, 9*(9), 1540. https://doi.org/10.3390/su9091540.

Levidow, L., & Raman, S. (2019). Metamorphosing waste as a resource: Scaling waste management by ecomodernist means. *Geoforum, 98*, 108–122. https://doi.org/10.1016/j.geoforum.2018.10.020.

Osburg, V.-S., Strack, M., & Toporowski, W. (2016). Consumer acceptance of Wood-Polymer Composites: A conjoint analytical approach with a focus on innovative and environmentally concerned consumers. *Journal of Cleaner Production, 110*, 180–190. https://doi.org/10.1016/j.jclepro.2015.04.086.

Piselli, A., Baxter, W., Simonato, M., Del Curto, B., & Aurisicchio, M. (2018). Development and evaluation of a methodology to integrate technical and sensorial properties in materials selection. *Materials and Design, 153*, 259–272. https://doi.org/10.1016/j.matdes.2018.04.081.

Reuter, M. A., van Schaik, A., Gutzmer, J., Bartie, N., & Abadías-Llamas, A. (2019). Challenges of the circular economy: A material, metallurgical, and product design perspective. *Annual Review of Materials Research*, *49*(1), 253–274. https://doi.org/10.1146/annurev-matsci-070218-010057.

SusCritMat. (2020). *EIT Raw Materials Academy, supported by the EU*. Available online: https://suscritmat.eu/. (Accessed January 9, 2020).

Wittstock, R., Pehlken, A., & Wark, M. (2016). Challenges in automotive fuel cells recycling. *Recycling*, *2016*(1), 343–364. https://doi.org/10.3390/recycling1030343.

Author Index

A

Abadías-Llamas, A., 227
Achzet, B., 53–54
Adachi, G., 54
Aggar, M., 123
Aggarwal, S., 30
Aghahosseini, A., 72
Agusdinata, D.B., 80
Agyeman, J., 209
Al Barazi, S., 157–158
Albertus, P., 82–83
Alexandrov, G., 2
Ali, S.H., 2, 109–110
Alman, D., 82–84
Alonso, E., 109–110
Alvarenga, R., 127–130
Amis, E.J., 133
Angerer, G., 66–67, 134, 169
Anghie, A., 99, 101
Anlauf, A., 154
Ardente, F., 126
Arent, D., 30
Arndt, N., 169
Ashfield, M., 31–34, 50–54
Auer, J., 65
Aurisicchio, M., 227
Aves Dias, P., 2
Axsen, J., 215–216, 215t
Azevedo, I.L., 30

B

Bach, V., 145–146, 178–179
Bahrudin, F.I., 227
Baier, M., 157–158
Bailey, S., 155
Bakker, S., 204
Banks, M., 123
Barr, R., 31–32, 42, 110–111, 135, 145–146
Bartie, N., 227
Barume, B., 182
Bassen, A., 2
Bastein, T., 32
Bates, A., 82–83
Bauer, C., 122, 124
Bauman, Z., 213
Baumann, M., 71–86
Baumann, M.J., 81
Baxter, W., 227

Baxter, W.L., 227
Bazilian, M., 2
Beason, J., 63
Bebbington, A., 14–15
Becker, E., 209–210
Behm, R.J., 82–83
Behrisch, J., 40–41
Belk, R.W., 212
Benighaus, C., 2
Benini, L., 93, 104, 126–129, 132–133, 135–136
Benson, P., 14
Berger, M., 124–126, 128–129, 131t
Bermudez, S., 170
Bernhardt, E.S., 13
Bertram, M., 40
Bessarabov, D., 102
Bhosale, S.V., 125–126
Bigum, M., 127
Blagoeva, D., 126–127
Blagoeva, D.T., 2
Blättel-Mink, B., 208–209
Bleicher, A., 2, 227
Bleischwitz, R., 214, 217
Blengini, G.A., 126–127
Bodin, J., 214–215, 215t
Bogdanoy, D., 72
Böhm, G., 214
Bonacker, T., 155
Bonatto Minella, C., 82–83
Bond, P., 103
Boonman, H., 32
Bosch, S., 66–67
Böschen, S., 50
Boudet, H.S., 215t
Bourdieu, P., 213
Bourgoin, A., 95
Boutt, D., 169
Bradley, D.C., 169
Brady, T., 130, 136
Brainard, J., 51–52
Brand, U., 209–210, 214, 217
Brandão, M., 121–122, 125–130, 131t, 133, 135
Branner, K., 63
Brehmer, E., 18
Brentrup, F., 135
Bréon, F.M., 129
Brereton, D., 115
Breyer, C., 94
Bridge, G., 156

Bringezu, S., 169
Brogaard, L., 127
Brunnengräber, A., 214
Bryant, R.L., 155
Buchholz, D., 74–75, 84
Buchholz, P., 31–32
Buijs, B., 121, 132–134
Bulach, W., 216–217
Burchart-Korol, D., 126, 128–129
Burritt, R.L., 114–115
Busch, J., 135
Bush, S.S., 96
Bustamante, M.L., 40–41
Bwenda, C., 160–161

C

Cai, G., 2
Cai, W., 30, 32
Campbell, C., 208, 213
Cao, W., 63
Capurro, F., 2
Carlisle, K., 84
Carvalho, M., 72–74
Castellani, V., 128–129, 136
Ceder, G., 154, 157–158
Celestin, B.L.N., 42
Chakrabarty, D., 92
Chandler, C., 31–32, 42, 110–111, 135, 145–146
Chang, H., 83–84
Chapman, A.J., 117
Chapman, B., 196
Chase, T., 31–32, 42, 110–111, 135, 145–146
Chen, C., 2
Chen, H., 215t
Chen, L., 82–83
Chen, L.-Y., 30, 32
Chen, W.-Q., 27–44
Cheung, G., 2
Child, M., 72
Choi, J., 31–32, 42, 110–111, 135, 145–146
Choi, S., 81–83
Choi, Y., 115
Christ, K.L., 114–115
Christensen, J., 82–83
Christensen, T., 127
Christoffersen, L., 31–32, 42, 110–111, 135, 145–146
Church, C., 154
Ciacci, L., 226
Cimprich, A., 92–93, 96, 100, 133, 145–146
Clagett, N., 13
Clark, N., 14

Clement, J., 169
Clowes, W., 98
Cobham, A., 96
Cocks, T., 98
Coe, N.M., 156
Cohen, D., 40
Collins, W., 129
Conde, M., 153–154
Cook, R., 183
Corcuera, S., 82–83
Corder, G.D., 109–110, 112–113, 115
Crawford, A., 154
Crossley, R.J., 63
Cumbers, A., 156
Cuvelier, J., 159–160

D

Datta, M.K., 82–84
Davies, P.J., 2
Davis, S.J., 30
Dawson, D.A., 135
de Almeida, A.T., 214
de Almeida, J., 127–130
de Koning, A., 126, 128–129, 136
De Koning, R., 159
De Lara Garcia, J.P.S., 62–63
de Schryver, A., 129–130, 131t
De Tomi, G., 179
De Ville, F., 94
Dees, D.W., 82–83
Defila, R., 209
Del Curto, B., 227
Delgado, J., 214
Deng, Y., 82–83
Dewulf, J., 126–127
Di Giulio, A., 208–209
Dicken, P., 156
Dicks, A.L., 109–110, 112–113
Diemant, T., 82–83
Dietrich, J., 123
Dietz, K., 153–155
Dietz, T., 214–215, 215t
Dietzenbacher, E., 2
Diniz Da Costa, J.C., 109–110, 112–113
Divekar, A., 182
Doganova, L., 95
Dominish, E., 92–94, 97–98, 100–102, 109–111
Doran, R., 214
Dorner, U., 178–179
Douglas, M.T., 213
Drielsma, J.A., 130, 136
Drnek, T., 130, 136

Du, X., 32, 135
Ducellier, B., 182
Duclos, B.S.J., 31–32
Duffer, H., 27–28
Dufo-López, R., 72–74
Dumit, J., 92

E

Eakin, H., 80
Eckelman, M.J., 36–40
Eggert, R.G., 42, 44
Eisenmenger, N., 53–54
Ellis, B.L., 82–83
Ellsworth-Krebs, K., 207
Elshkaki, A., 113–114
Elshout, P.M.F., 110–111
Elwert, T., 216–217
Emel, J., 98–99, 101
Engels, B., 153–155
Erb, K.H., 53–54
Erdmann, L., 66–67, 109–110, 132, 134, 153, 169
Ernstson, H., 92
Eroglu, D., 82–85
Escobar, A., 155
Espinal, L., 133
Espinoza, L.T., 40–41
Estornés, J., 82–83
Evans, D., 214, 217–218
Evans, D.M., 208–209, 212

F

Faber, B., 159
Fagan, A., 141–142
Faria, R., 214
Fell, H.-J., 94
Ferguson, N., 96
Ferry, E.E., 91, 93–94
Field, F.R., 109–110
Fine, B., 211–212
Finkbeiner, M., 124–126, 128–129, 131*t*, 178–179
Fischer, D., 208–209
Fisher, S., 17–18
Fitzpatrick, P., 14–15
Fladager, G., 19
Florin, N., 92–94, 97–98, 100–102
Fonseca, A., 14–15
Forslund, D., 103
Fortier, S.M., 51–52
Fowai, I., 63
Frankel, T.C., 154
Franken, G., 169–184
Fraunhofer, I.S.I., 92, 96

Freire, F., 214
Frenzel, M., 31–32
Friedman, L., 12
Frischknecht, R., 127–128
Fu, X., 154, 157–158
Fuglestvedt, J., 129

G

Gaines, L., 81
Gallagher, K.G., 82–85
Gambogi, J., 51–52
Gao, X., 82–83
Garche, J., 83–84
García-Olivares, A., 217
Gauß, R., 53
Gaustad, G., 40–41
Gaustad, G.G., 154, 157–158
Ge, J.-P., 30, 32
Geenen, S., 159–160, 162
Gemechu, E.D., 92–93, 96, 100, 133, 145–146
Geng, J., 215*t*
Ghoreishi-Madiseh, S.A., 115
Giegrich, J., 127–128, 130, 131*t*, 132
Gilbert, P.R., 91–105
Gingrich, S., 53–54
Giurco, D., 40–41, 92–94, 97–98, 100–102, 109–111
Giurco, D.P., 109–110, 112–113
Glöser-Chahoud, S., 62–63
Gobbers, E., 159–160
Goddin, J.R.J., 187–197
Goebel, S., 82–85
Goedkoop, M., 129–130, 131*t*
Goffé, B., 169
Goldmann, D., 216–217
Gonzalez, G., 65
Goodland, R., 14
Goodman, M.K., 216–217
Goonan, T.G., 40–41
Góralczyk, M., 126, 132
Gordon, R.B., 40
Graedel, T.E., 31–34, 40–42, 109–111, 132, 134–135, 145–146, 226
Grande, L., 84–85
Green, M.L., 133
Greszler, T., 82–85
Gronow, J., 212
Gronwald, V., 179
Grosjean, C., 109–110
Grunwald, A., 153–154
Guillaume, B., 31–32
Guinée, J., 128–130, 131*t*

Guinée, J.B., 126, 128–129, 131*t*, 136
Gulagi, A., 72
Gulakowski, J., 84
Gunn, G., 40–41
Gutfleisch, O., 53
Güth, K., 53
Gutstein, D., 95–96
Gutzmer, J., 31–32, 227

H

Ha, S., 82–83
Haas, T., 214
Haberl, H., 53–54
Habert, G., 31–32, 92–94
Habib, K., 31–32, 92–94, 96–98, 104
Hagelüken, C., 2, 31–32, 214, 217
Hamelin, L., 97–98
Hao, H., 157–158
Harper, E.M., 31–34, 41–42
Harvey, L.D.D., 109–110
Hassani, F.P., 115
Hassoun, J., 84–85
Hatayama, H., 27–28
Hatch, G., 19–20
Hatfeld-Dodds, S., 169
Hauck, M., 32
Haus, R., 156
Hauschild, M., 131*t*
Hayes, S.M., 31–32, 141, 145–146
Heeren, N., 31–32, 92–94
Hefferman, T., 21
Heijungs, R., 129–130, 131*t*
Helbig, C., 92–93, 96, 100, 133, 145–146
Hellweg, S., 127–128, 169
Henderson, J., 156
Hendriks, H.W.M., 127–128
Hendriks, J.A., 127–128
Hess, M., 156
Hesse, H.C., 74
Hilson, A., 183
Hilson, G., 183
Hinton, E.D., 216–217
Hobson, P., 101–102
Hoffmann, V.H., 2
Hofmann, H., 214
Hofmann, M., 214
Hollins, O., 92, 96
Hönke, J., 159–160
Höök, M., 113–114
Hool, A., 214
Hossain, K., 99
Hossain, M.S., 2

Hu, T., 101–102
Huang, J., 129
Huber, M., 93
Huber, M.T., 98–99, 101
Hughes, T., 15
Huijbregts, M., 130
Huijbregts, M.A.J., 110–111, 127–128
Hume, N., 98
Hummel, D., 209–210
Hummen, T., 66–67, 134, 153, 169
Hungerbühler, K., 127–128
Huppes, G., 126, 128–129, 136
Huston, D.L., 27–28
Hwang, J.-Y., 82–84
Hynan, P., 72–74
Hynek, S.A., 169

I

Ihle, C.F., 114–115
Imanaka, N., 54
Imbusch, P., 155
Ioannidou, D., 31–32, 92–94
Isherwood, B., 213
Ishihara, K.N., 109–110, 112–113

J

Jacka, J.K., 2
Jackson, T., 212
Jaffe, R.L., 40–41
Jahanshahi, S., 111
Jahn, T., 209–210
Jamasmie, C., 98
Janský, P., 96
Jansson, J., 214–215, 215*t*
Jena, A., 83–84
Jin, Y., 31–32
Jochens, H., 169
Jonas, H., 149–150
Jossen, A., 74
Justesen, L., 91

K

Kagawa, S., 92–93, 95–96
Kaikkonen, L., 76
Kalverkamp, M., 227
Kama, K., 91, 93–95, 100–101
Kantor, M., 96–97
Kara, S., 32
Karim, K.S., 133
Katz-Lavigne, S., 154
Kaufmann-Hayoz, R., 209

Kavlak, G., 32, 40–41
Kemp, D., 2, 14–15, 112–113
Khan, A.F., 2
Kickler, K., 169–184
Kim, J., 31–32
Kim, J.G., 82–83
Kinchy, A., 15–16
Kirchain, R.E., 109–110
Kirchner, N., 44
Kirsch, S., 14
Kirsh, S., 15–16
Kleijn, R., 66–67
Kleinhückelkotten, S., 209
Klinglmair, M., 121–122, 125–130, 131*t*, 133, 135
Knoeri, C., 135
Knöfel, S., 125
Kobrin, S.J., 96
Koch, B., 129, 131*t*, 133, 135, 141–151
Kohout, M., 20
Kojola, E., 2
Kollmuss, A., 209
Kondo, Y., 92–93, 95–96
Konitzer, D.G., 31–32
Kornberger, M., 91
Kozinsky, B., 82–83
Kracht, W., 114–115
Krause, B., 159
Krausmann, F., 53–54
Kruczek, M., 126, 128–129
Kucevic, D., 74
Kuhlmann, H., 135
Kühn, A., 62–63
Kühnel, K., 178–179
Kuikka, S., 76
Kulionis, V., 2
Kullik, J., 31–32
Kumta, P.N., 82–84
Kurzweil, P., 83–84
Küsters, J., 135
Kwon, O., 82–83

L

Labay, K., 169
Lagos, G., 40
Lammel, J., 135
Lang, D.J., 214, 217
Langkau, S., 2, 66–67, 134, 153, 169
Lapko, Y., 216–217
Larrea, C., 170
Latour, B., 15
Lauwigi, C., 127–128, 130, 131*t*, 132
Law, Y.-H., 42

Lawhon, M., 92
Le Billon, P., 153–154
Lebedeva, N., 74
Lèbre, É., 112–113
Lederer, G.W., 51–52
Lee, J.C.K., 42
Lee, Y.-J., 84–85
Lentz, A., 215–216
Levidow, L., 2
Lewis, N.S., 30
Li, F., 14
Li, J., 82–83
Li, J.S., 27–44
Li, N., 27–44
Li, R., 82–83
Li, T., 82–83
Li, W., 215*t*
Liang, J., 82–83
Licht, C., 32
Lidia, C., 42
Liebich, A., 127–128, 130, 131*t*, 132
Liedtke, M., 157–158
Liedtke, R., 82–83
Limbert, M., 91, 93–94
Lin, X.-M., 82–83
Linarelli, J., 99, 101
Liu, R.-S., 83–84
Liu, S., 2
Liu, T., 216–217
Liu, W., 80
Liu, Z., 157–158
Lohmann, T., 82–83
Long, R., 215*t*
Lossin, A., 60
Lotoskyy, M., 102
Løvik, A.N., 2, 31–32
Lu, W., 82–83
Luis, A., 121, 132–134
Luo, J., 82–83
Lyman, R., 20
Lynch, M., 171–172, 175, 178

M

Mabanza, B., 102–105
Madlool, N.A., 2
Madsen, A.K., 91
Maennling, N., 177
Mahamba, F., 98
Månberger, A., 132
Mancini, L., 93, 104, 126–128, 132–133, 135
Manivannan, A., 82–84
Mao, Y., 143

Marin, T., 179
Marmier, A., 2
Marques, P., 214
Marscheider-Weidemann, F., 66–67, 134, 153, 169
Martinez-Alier, J., 2, 155
Masi, D., 216–217
Mason, L., 40–41
Masson-Delmotte, V., 71
Mathe, M., 102
Mathias, M., 82–85
Matthysen, K., 183
Mc Quilken, J., 183
McAllister, M., 14–15
McCullough, E.A., 31–32, 51–52, 141, 145–146
McGlynn, B.L., 13
McGugan, M., 63
McLellan, B., 92–94, 97–98, 100–102, 113–114
McLellan, B.C., 109–118
McManus, M.C., 169
McNerney, J., 40–41
Meesala, Y., 83–84
Meinzer, M., 96
Menictas, C., 82–83
Menoufi, K., 124–125, 127–128
Mernagh, T.P., 27–28
Metayer, M., 94
Miao, Y., 72–74
Michelsen, G., 208–209
Miranda, P.H., 109–110
Mishnaevrsky, L., 63
Mistry, M., 130, 136
Mitchell, P., 183
Mitchell, T., 15
Mizuno, F., 84
Mohamed, A., 84
Mohr, M., 74–75, 81
Mohtadi, R., 84
Moreno-Ruiz, E., 122, 124
Morris, L., 203
Moser, S., 209
Mosig, J., 125
Mossallam, M., 101
Moura, P., 214
Mouritsen, J., 91
Mouton, K., 102
Mudd, G., 40–41
Mudd, G.M., 113–115
Mukherjee, S., 82–83
Müller, D.B., 32, 74–75, 79, 153–154
Mulligan, D., 115
Muniesa, F., 91, 95
Munk, L.A., 169
Murakami, S., 113–114

Musamba, J., 162
Myhre, G., 129
Myung, S.-T., 82–84

N

Näher, U., 157–158, 182
Nakajima, K., 92–93, 95–96
Nansai, K., 92–93, 95–96, 109–111
Nassar, N.T., 31–34, 40–42, 51–52
Nathan, S., 84
Nativel, C., 156
Nauman, T., 19
Nazar, L.F., 82–83
Neil, D., 115
Nelly, D.S., 42
Nemery, B., 2
Neumann, M., 182
Ng, E., 54
Ngoie, G.T., 159
Ni, O., 65
Nita, V., 126–127
Niu, X., 57, 57*t*
Norgate, T., 111
North, B., 102
Nuss, P., 31–34
Nuur, C., 216–217
Nye, D., 15
Nygård, H., 76

O

Ober, J.A., 40
Oberhaus, D., 194
Oberle, B., 169
Oelerich, W., 82–85
Ogata, S., 92–94, 109–110, 117–118
Okatz, J., 2
Oliveira, I., 127–130
Olivetti, E.A., 154, 157–158
Ölz, S., 44
Ombeni, A., 182
Omeje, K., 159
Onstad, E., 101–102
Orgil, K., 204
Origliasso, C., 179
Ortiz, H., 95
Osburg, V.-S., 226–227
Oswald, I., 134
Osychenko, O., 217
Otto, J.P., 31–32
Overland, I., 2, 100, 214
Owen, J.R., 2, 14–15, 112–113
Oyewo, A.S., 72

P

Paillard, E., 84–85
Pandey, A.K., 2
Pant, R., 128–129, 136
Park, J.-B., 84–85
Pasaoglu, G., 72–74
Passarini, F., 226
Passerini, S., 74–75, 84
Pasupathi, S., 102
Paterson, F., 95
Pavel, C.C., 2
Peake, S., 71
Pehlken, A., 121–136, 131t, 227
Peiró, L.T., 32
Peña Cruz, A., 81–82
Peñaherrera, F., 129, 131t, 133, 135
Pennington, D., 126, 132
Perrin, M., 109–110
Peters, J., 71–86
Peters, J.F., 74–75, 79, 81–84
Petersen, H.N., 63
Peterson, C., 19
Pfister, H.-R., 214
Phadke, R., 2, 11–23
Philip, N., 36–40
Pina-Stranger, Á., 95
Piselli, A., 227
Poggi, P., 109–110
Pollet, B.G., 102
Polonsky, A., 84
Pörtner, H.O., 71
Potting, J., 131t
Potts, J., 171–172, 175, 178
Prakash, J., 82–83
Prause, L., 153–164
Prior, T., 40–41
Purnell, P., 135
Purvis, B., 143
Pye, O., 156

Q

Qiu, L., 2

R

Radley, B., 2, 161–162
Raeymaekers, T., 162
Ragnarsdóttir, K.V., 40
Rahim, N.A., 2
Rajak, D., 101
Ram, M., 72
Raman, S., 15, 208
Ranjith, P.G., 125–126

Recchioni, M., 132
Reck, B.K., 31–34, 109–110, 226
Reid, L., 207
Reinhard, J., 122, 124
Reinhardt, J., 127–128, 130, 131t, 132
Reisch, L.A., 213
Reller, A., 31–34, 50–54
Renn, O., 216–217
Rennie, C., 31–34, 50–54
Restrepo, E., 32
Reuter, M.A., 31–32, 227
Revell, R., 135
Rezvani, Z., 214–215, 215t
Richardson, T., 91, 93
Richins, M.L., 213
Riemann, A., 31–32
Riemke, R.L., 2
Rietveld, E., 32
Ritardo, D.T., 60
Ritchken, E., 101–102
Ritthoff, M., 2, 135
Roberts, D., 71
Robinson, D., 143
Roelich, K., 135
Rogers, E.M., 215
Romero, H., 80
Rosenau-Tornow, D., 31–32
Ross, A., 159
Ross, M.R.V., 13
Routledge, P., 156
Rubbers, B., 159–160
Ruhland, K., 125
Rule, T.A., 2
Russell-Vaccari, A.J., 130, 136

S

Saha, P., 82–84
Sainsbury, P., 103
Sala, S., 93, 104, 121–122, 125–130, 131t,
    132–133, 135–136
Salomon, M., 99, 101
Samadi, S., 2, 135
Sanchez de la Sierra, R., 159
Sanchez-Carrera, R.S., 82–83
Sanderson, H., 97–98, 157
Sanne, C., 209
Saugstad, J., 146–147
Schandl, H., 169
Schebek, L., 74–75, 79, 153–154
Schelly, C., 215–216, 215t
Scheyder, E., 12–13
Schimpe, M., 74
Schipper, B., 66–67

Schlemmer, E., 101
Schmid, D., 216–217
Schneider, L., 124–126, 128–129, 131*t*, 135
Schneider, N.F., 208
Schödwell, B., 128–130, 131*t*
Schouten, P., 183
Schrijvers, D., 145–146
Schubel, P.J., 63
Schularick, M., 96
Schüler, D., 216–217
Schüler-Hainsch, E., 125
Schuppert, N., 82–83
Schütte, P., 157–158, 178–179, 181–182
Scott, J.B., 141–142
Seccatore, J., 179
Sellin, G., 216–217
Selvaraj, J., 2
Senior, A., 27–28
Shaker, R.R., 144
Shanbhag, S., 84
Shaner, M., 30
Shaw, L.L., 82–83
Sheridan, R., 53
Shigetomi, Y., 92–93, 95–96
Shindell, D., 129
Shove, E., 209, 217
Shukla, P., 71
Sievers, H., 121, 132–134
Siles-Brugge, G., 94
Silver, J., 92
Simmons, J., 31–34, 50–54
Simonato, M., 227
Sims, R., 44
Sinsel, S.R., 2
Skea, J., 71
Skirrow, R.G., 27–28
Slater, D., 211, 211*np*
Smale, R., 2
Smil, V., 28
Smith Rolston, J., 15
Smith, J., 15–16
Smith, N.M., 2
Soentgen, J., 50
Solé, J., 217
Son, B., 82–83
Song, J., 157–158
Sonnberger, M., 207–218, 215*t*
Sonnemann, G., 31–32, 92–94, 133, 145–146
Sørensen, B.F., 63
Sornarajah, M., 99, 101
Soukup, O., 2, 135
Sovacool, B.K., 2, 154, 215*t*

Spaargaren, G., 2
Spittaels, S., 183
Sprecher, B., 66–67
Spriensma, R., 129–130, 131*t*
Stam, G., 110–111
Steinmann, Z.J.N., 110–111
Stenqvist, B., 132
Stern, P.C., 209, 214–215, 215*t*
Steubing, B., 122, 124
Storrow, B., 19
Strack, M., 226–227
Struijs, J., 129–130, 131*t*
Stückrad, S., 2
Styve, M.D., 103
Su, B., 216–217
Suh, S., 92–93, 95–96
Sun, K., 63
Sun, Q., 82–83
Sun, X., 82–83, 157–158
Sun, Y.-K., 82–85
Svensson, H., 63
Swaffield, J., 214, 217–218
Swart, G., 102

T
Tahara, K., 27–28
Tamura, S., 54
Tan, Z., 63
Tarvydas, D., 74
Tercero Espinoza, L., 66–67, 134, 153, 169
Tercero Espinoza, L.A., 2
Tercero, E., 121, 132–134
Tercero, E.L.A., 62–63
Teubler, J., 2, 135
Tezuka, T., 92–94, 109–110, 117–118
Tharumarajah, A., 125–126
Thierry, D.P., 42
Thomas, H., 2
Thorenz, A., 92–93, 96, 100, 133, 145–146
Thorne, J.P., 27–28
Tilton, J.E., 40
Tokimatsu, K., 113–114
Toledo, P., 177
Tony, K., 42
Toporowski, W., 226–227
Torres de Matos, C., 126–127
Trancik, J.E., 40–41
Traversa, E., 133
Trianni, A., 216–217
Truelove, H.B., 2
Tsing, A., 95
Tsiropoulos, I., 74

Tuazon, D., 115
Tübke, J., 71
Tuma, A., 92–93, 96, 100, 133, 145–146
Turley, L., 169–184
Turner, G.M., 40
TyreeHageman, J., 215–216

V

Vaalma, C., 74–75
Vaca, F.P., 121–136
Valenta, R.K., 112–113
van der Merwe, F., 102
van der Voet, E., 121, 134–135
Van Exter, P., 66–67
van Harmelen, T., 32
van Niekerk, F., 102
van Oers, L., 126, 128–130, 131t, 136
van Schaik, A., 227
van Vliet, B., 2
van Wijk, J.M., 76
van Zelm, R., 129–130, 131t
Veblen, T., 213
Veiga, M., 179
Velasquez, R., 65
Velikokhatnyi, O.I., 82–84
Venesjärvi, R., 76
Ventus, A., 64t
Verones, F., 110–111
Vetter, S., 157–158
Vidal, O., 169
Vidal-Legaz, B., 126–129, 136
Viebahn, P., 2, 135
Vieira, M.D.M., 110–111
Villalba, G., 32
Vincent, H., 42
Vogel, C., 162
von Jouanne, A., 72–74
Von Kessel, I., 191–192
Voora, V., 170
Vosloo, M., 102

W

Wachtmeister, H., 113–114
Wäger, P., 2, 31–32
Wäger, P.A., 214, 217
Wagner, M., 31–32
Walton, A., 53
Wang, D., 82–83
Wang, G., 81–83
Wang, L., 2
Wang, P., 27–44

Wang, Q., 216–217
Wang, X., 74–75, 81
Warde, A., 208, 212
Wark, M., 227
Watari, T., 92–94, 97–98, 100–102, 109–111,
    117–118
Weber, S., 81
Weidema, B., 122, 124
Weihed, P., 130, 136
Weil, M., 71–86, 153–154
Welch, D., 214, 217–218
Welker, M., 15–16
Wen, Z., 42
Wenban-Smith, M., 171–172, 175,
    178
Wentland, A., 214
Wenzel, H., 31–32, 92–94, 96–98, 104
Wernet, G., 122, 124
Weszkalnys, R., 91, 93
Whitacre, J.F., 84
Whitehouse, D., 103
Wiesen, K., 2, 127–128, 135
Wilburn, D.R., 40–41
Williams, M., 102
Winner, L., 15
Wirges, M., 127–128
Wissen, M., 209–210, 214, 217
Wiswede, G., 214
Wittmer, D., 214, 217
Wittstock, R., 227
Wolske, K.S., 214–215, 215t
Worrall, R., 115
Wu, Q., 82–83
Wu, Y., 2
Wuest, T., 227

X

Xie, Y., 63
Xue, B., 2

Y

Yamasue, E., 109–111
Yang, X., 82–83
Yang, Y., 53
Yasuoka, R., 113–114
Yellishetty, M., 125–126
Yeung, H.W.-C., 156
Yokochi, A., 72–74
Young, S.B., 92–93, 96, 100, 133
Yuan, C., 82–83
Yusoff, K., 14

Z

Zarnekow, R., 128–130, 131*t*
Zepf, V., 31–34, 49–68
Zhai, P., 71
Zhang, J., 63

Zhao, F., 2, 157–158
Zhao-Karger, Z., 82–83
Ziemann, S., 72, 74–75, 79, 85, 153–154
Zubi, G., 72–74

Subject Index

Note: Page numbers followed by *f* indicate figures and *t* indicate tables.

A

Abiotic Depletion Potential (ADP), 128–129
Abiotic resource depletion, 126, 130
ACAES. *See* Adiabatic compressed air energy storage (ACAES)
Accelerated Metallurgy project, 204
Acid mine drainage (AMD), 115
Adiabatic compressed air energy storage (ACAES), 71–72, 73*f*
Advanced lithium-ion batteries (a-LIB), 84–85
AHIB. *See* Aqueous hybrid-ion batteries (AHIB)
a-LIB. *See* Advanced lithium-ion batteries (a-LIB)
Aqueous hybrid-ion batteries (AHIB), 84
Artisanal and small-scale mining (ASM)
    challenges, 182–183
    limited market penetration, 179
Asynchronous generators (ASG), 62–63
Autocatalysis technologies, 36–40
Avoided Environmental Burden (AEB), 134

B

Bilateral Investment Treaty (BIT), 99, 101
    South Africa and Congo, 101
Biofuel, 30–31
Biomass, 55–56
    fermentation plants, 55
    thermal power plants, 55
BlueMAP scenario, 94
Brundtland Report, 143–144, 146, 148–151

C

Cadmium-telluride (CdTe), 60
Capitalization, 95
Circular economy, 226–227
    access-based models, 179
    butterfly diagram, 191, 192*f*
    environmental benefits, 179
    material substitution, 199–204
    nonmaterial resources, 53–54
    product design and materials recovery, 176–177
    recycling technologies, 53, 76, 226–227
    reducing supply risks, 172–175
    resource supply and competition, 172–178
Circular economy approach, 6–7
Circularity of a product, 177
Clean Power Plan, 11–12

Coal, 30–31
    mining, 116
Cobalt, 85, 97–99
    extraction, 156–157
    demand, 78–82, 79*f*
Cobalt mining, in DRC
    Democratic Republic of Congo (DRC), 6, 154
    extractive industry *vs.* post-colonial host states, 96–97
    global cobalt production, 157*f*
    international NGOs *vs.* local host communities, strategic alliance, 163
Concentrated solar power (CSP), 59, 61–62
Conflicts
    cultural, 155
    definition, 155
    minerals, 4, 42, 171
    mining (*see* Mining conflicts)
    resources, 4, 141–142, 145
    social, 42
Consumption
    environmental motivation, 212
    extraordinary, 212
    function of, 213
    inconspicuous, 212
    ordinary, 212
    patterns, 209–210, 217
    society-nature relations, 209–210, 210*f*
    sociology of, 207–218
    symbolic meaning, 209–210, 214, 218
Copper-indium-gallium-diselenide (CIGS), 60
CRAFT standard, 176
Creeping regulatory expropriation, 96–97
Criticality, 4–5, 49, 51–53
    assessment methodologies, 92–93, 96
    description, 93
    material, 31–32, 42
    matrix, 92–93
    mineral criticality, 109–110
        evaluation technique
        see Life cycle assessment (LCA)
Criticality Weighted Abiotic Depletion Potential (CWADP), 129
Critical minerals mining, 11, 40, 42
    Alaska, 17–18
    social license to operate, 21–22
    development, 14

Critical minerals mining *(Continued)*
  environmental impacts, 42
  geopolitics, 22
  labor transitions, role of, 23
  political and social challenges, 11
  Texas, 19–21
  Wyoming, 18–19
Critical raw materials (CRMs), 92–93, 141
  critical elements, 21–22, 52, 52f
  critical materials, 4
  critical resources, 4, 141–142, 145
  substitution of (*see* Substitution)
CSP. *See* Concentrated solar power (CSP)
Cumulated Energy Demand (CED), 126–128, 130, 133
Cumulated Material Demand (CMD), 127–128

D

Decarbonization, 91–92, 97–98
Democratic Republic of Congo (DRC)
  AVZ minerals, 100–101
  lithium, 100
  royalty rates, 98
  strategic mineral resources, 98
Doubly-fed asynchronous induction generator
  (DFIG), 62–63
Drive train system (DTS), 62
DTS. *See* Drive train system (DTS)

E

ECEs. *See* Energy-critical elements (ECEs)
Eco-Indicator 99 method
  endpoint, 130
  midpoint, 129
Economic Commission for Latin America and the
  Carribean (ECLAC), 184
EDIP. *See* Environmental Development of Industrial
  Products (EDIP)
Electric vehicle (EV), 7, 74, 92, 97–98, 153, 207, 214
  adopters, 215–216
  batteries, 74–75
  case, 214–216
  cobalt, in EV production, 155–156, 160–161
  electricity consumption, 212
  embodied energy, 177f
  mobile application of batteries, 74–75
  permanent magnets, 180
  price volatility of key elements, 177f
  production and consumption, 156–158
Electrification of transport sector, 153
  lithium-ion batteries, 153–154
Endpoint methods, 129–130
Energy-critical elements (ECEs), 31–32, 33f

Energy
  demand, 28
  related materials, 133
  related tools, 31
  source, 30–31
  storage demand, 72
  transition, 28–31, 29f, 163–164
    circular economy, role of, 187–196
    electrification of transport sector, 153–154
Environmental awareness, 209, 215t
Environmental Development of Industrial Products
  (EDIP), 127–128, 130
  EDIP2003 methodology, 133
Environmental Impact Assessment (EIA), 111–112
Environmental impacts, 42
  energy and, 113–114
  lifecycle assessment, 110–111
  legacy mines, 115
  local perspective, 111–113
  emissions at demand side, 117
  mines, 114–116
  processing, extraction, smelting, and refining,
    116–117
  waste and recycling, 117–118
Environmental justice, 15–16, 20–22
Environmental Management Plan (EMP), 112
Ethical dilemma, 148–150
Expropriation
  creeping regulatory, 96–97
  taxation rates, 96–97
Extractive Industries Transparency Initiative (EITI),
  170, 184
Extractive industry investment
  capitalization, 95
  discount rate, 95

F

Flat plate solar collectors, 61
Francis turbines, 57–58
Future generations
  consideration, 149
  needs of, 148–149
  Parfit's paradox, 146–148
  precautions for, 149–150

G

General Land Office (GLO), 20
GLO. *See* General Land Office (GLO)
Global energy transition, 28–31
  element basis, 31
  energy-critical elements, 31–32
  global challenges

by-production constraints, 41–42
material needs, 44
metal supply, 43
metal system, 44
mining operation time, 40–41
physical scarcity, 40
source, 30–31
sustainable material cycle in, 32–40
Global material extraction data, 53–54
Global production networks (GPN), 156, 162–163
Global Warming Potential (GWP), 129
Greenhouse gases (GHG), 121–122
Green New Deal, 12

H

Heat pipe solar collectors, 61
Herfindahl-Hirschman Index (HHI), 187
Hydroelectric power plants, 56–58, 57t
Hydrogen fluoride (HF), 81–82
Hydrogen South Africa (HySA) strategy, 102

I

IEA. *See* International Energy Agency (IEA)
Impact assessment
    material resource consumption, 131t
Imported nonfuel materials, 92–93
International Council of Mining and Metals
        (ICMM), 171, 174, 178–179, 182
International Energy Agency (IEA), 30, 94

K

Kaplan turbine, 57–58

L

LCA. *See* Life cycle assessment (LCA)
Lead-acid batteries (PbA), 81
LIB. *See* Lithium-ion batteries (LIB)
LIB-LCO. *See* Lithium cobalt oxide (LIB-LCO)
LIB-LFP. *See* Lithium-iron-phosphate batteries
        (LIB-LFP)
LIB-NCA. *See* Lithium nickel cobalt aluminum
        oxide batteries (LIB-NCA)
Life cycle assessment (LCA), 4–5, 110–111, 113,
        121–122, 170, 224–225
    ecoinvent, 122
    endpoint methods, 129–130
    goal and scope definition, 124
    impact assessment, 124, 125f, 126–127, 131f
    indicator selection, role of, 126–127
    interpretation, 124
    inventory analysis, 124
    ISO 14040 and 14044 standards, 123

midpoint methods, 128–130
    reliability, 122
    resource accounting methods, 127–128
    social life-cycle assessment (SLCA), 122, 133
Life cycle impact assessments (LCIA), 125–126,
        132–133
Lithium, 85, 100
    demand, 78–80, 79–80f
Lithium cobalt oxide (LIB-LCO), 72–74, 73t
    type battery, 75–76
Lithium-ion batteries (LIB), 72–74, 73t, 81–85,
        153–154
    drawbacks of, 81–82
    waste and recycling, 117–118
Lithium-iron-phosphate batteries (LIB-LFP), 73t,
        74, 79–80
Lithium nickel cobalt aluminum oxide batteries
        (LIB-NCA), 73t, 74
Lithium-oxygen (Li-O2)/lithium-air (Li-air)
        batteries, 84–85
Lithium-sulfur (Li-S) batteries, 84–85
Local environmental impact perspective, 111–113

M

Magnesium-ion batteries (MIB), 84
Material flow analysis (MFA), 32–33, 44
Materials Genome Initiative, 204
Metals
    battery-related metals, 34, 35f
    supply, 43
    photovoltaic (PV)-related metals, 34–36, 37f
    platinum group metals (PGMs), 36–40,
        101–103
    vehicle-related (automotive catalyst) metals,
        36–40, 39f
    wind/motor-related metals, 36
Metal-energy nexus, 28–32
MFA. *See* Material flow analysis (MFA)
MIB. *See* Magnesium-ion batteries (MIB)
Mineral-energy nexus framework, 3–4
Mineral Resource Depletion (MRD) impact
        category, 129
Mineral trade and geopolitics, 42
Mining conflicts
    artisanal miners *vs.* mining companies, 159
    bad labor conditions, 162–163
    critical collaborations, 15
    due diligence campaigns, 161–163
    eviction of artisanal miners, 162
    host local communities *vs.* industrial mining
        companies, 159–160
    human rights violations, 162–163

Mining governance
  due diligence regulation, 153–155, 161–163, 171, 180, 183
  Bilateral investment treaty, 99
Mining operations
  environmental concerns
    atmospheric emissions, 115–116
    contamination/pollution, 19, 22, 115
    dust emissions, 115–116
    landform change, 116
    water management, 114–115
  finance, 91–92
  mining process, 40–41
  time, 40–41
Modified HHI indexes, 176*f*
Monopoly of supply, 188
  of key elements used in electric vehicles, 188*f*
Moral obligations, 146–151

N

NCA-type battery, 78–79
Neodymium, 62–63
Nickel, 85
  demand, 78–79, 79–80*f*
Nickel manganese cobalt batteries (NMC), 73*t*, 74
  NMC 111, 76–77
  NMC 622, 77–78, 77*f*
  NMC 811, 78
NMC. *See* Nickel manganese cobalt batteries (NMC)

O

Oil (gas), 30–31
Oxygen cathode, 84–85

P

Pelton turbine, 57–58
Permanent magnet generator (PMG), 62–63
PHES. *See* Pumped hydroelectric energy storage (PHES)
Photovoltaics (PV), 59–61
  installations, 7
  panels, waste and recycling, 117–118
  systems, 212, 214
    adopters, 215–216
    rooftop, 207
PLB. *See* Postlithium battery (PLB)
Policy Potential Index (PPI), 95–96, 104
Political ecology, 155–156, 162
Political risk
  assessment, 95–97

calculation, 95–96
Confiscation, expropriation, and nationalization (CEN) risk, 96–97
metrics, 91–92
valuation and critical raw materials, 95–97
Post-colonial, 91–92, 96–97, 99–101, 104–105
Postlithium battery (PLB), 84
  recycling and sustainability, 81–85
Power relations, 155–156, 163
Price volatility, 170–171, 174, 177*f*
Process substitution, 203
Product design, 227
Product substitution, 202–203
Pumped hydroelectric energy storage (PHES), 71–72, 73*f*
Pumped-storage reservoir, 58

R

Rare earth elements (REE), 49–50, 54
Raw Materials Information System (RMIS), 126–127
REACH regulation, 142–143
ReCiPe method, 129–130
REE. *See* Rare earth elements (REE)
Renewable energy, 50–51, 113, 117, 121, 134–135, 169
  biomass, 55–56
  criticality aspect, 51–53
  critical resources, 49–54
  geothermal energy, 65–66
  indicators and relationship to material demands, 132–133
  material resources, 53–66
  materials demand for, 61–62, 66–68
  reservoir type hydroelectric power plants, 56–58
  run-of-river hydroelectric power plants, 56–58
  water, 56–59
  wind energy, 62–65
Resource accounting methods (RAM), 127–128
Resource efficiency, 121–122
Resource imaginaries, 94, 96, 103
Resource indicators, in LCA, 126
Resource-making, 91, 95
Resource nationalism, 95–97
Resourceness, 93
Responsible
  mining, 5–6, 14–16, 22–23
  Responsible Cobalt Initiative, 161–163
  Responsible Minerals Initiative (RMI), 176, 180–181
RMI. *See* Responsible Minerals Initiative (RMI)
Run-of-river hydroelectric power plants, 56–58

S

Science and technology studies (STS), 15–16, 91, 94
Sea current power plants, 56
Sharing economy, 176
SIB. *See* Sodium-ion batteries (SIB)
Social Impact Assessments (SIAs), 111–112
Social license to operate, 21–22
Social movements, 153–154, 156, 164
Society-nature relations, 209–210, 210f
Sodium-ion batteries (SIB), 84
Solar cells, 54–55
Solar energy production, 59–62
Solar power tower plants, 62
Solar radiation, 72
Solar thermal energy (ST), 59, 61
Solid-state batteries (SSB), 83–84
SSB. *See* Solid-state batteries (SSB)
Stationary battery systems, 71–72
   advantage, 81–82
   future demand, for energy storage, 72
   lithium-ion batteries (LIB), 72–74
     recycling and sustainability, 81–85
   potential resource demand, calculation of, 74–80
     LCO-type battery, 75–76
     LFP-type battery, 79
     NCA-type battery, 78–79
     NMC111-type batteries, 76–77
     NMC 622-type battery, 77–78
     NMC 811-type battery, 78
Stoffgeschichten approach, 50
Subsoil social theory, 14
Substitution
   barriers/challenges to, 205
   new technology, 203–204
   process substitution, 203
   product substitution, 202–203
   rare-earth elements, 200, 202f
   substance substitution, 199–204
   transition, 214
Supply risk, 146
Sustainability, 142–144
   development, 143–144, 146
   initiatives, in mining sector, 172, 172–174t
   sustainable development goals (SDGs), 170 (*see also* Voluntary sustainability initiatives (VSIs))

Sustainable consumption, 209. *See also* Consumption
System of provision approach, 211–212

T

Technologies of imagination, 94
Tidal power plant, 58–59
   sea currents and tidal movements, 59
Total material requirement (TMR), 110–111
Traceability, 178
Transnational activism, 156

U

US critical minerals policy, 11–13

V

Valuation studies, 91
Vanadium production, 34
Voluntary sustainability initiatives (VSIs), 171
   assessing and categorizing, 174
   assurance systems, 178
   categories and issues, 176f
   compliance with regulations, 180
   cost of certification, 183
   deviant and illegal behavior, 182
   harmonization of standard requirements, 179–180
   mining and mineral supply chains transparency, 180–181
   reporting and performance, 182
   sustainable mining initiatives, 180
   sustainability requirements, 176–177
VSIs. *See* Voluntary sustainability initiatives (VSIs)

W

Water energy, 50–51, 56–59, 64t
Wind energy, 30–31, 50–51, 62–65, 64t
Wind radiation, 72
Wind turbine, 49–50, 54–55
   motor production, 36, 38f
Wind turbine generators (WTG), 62–63, 64t
   ASG-type, 63
   three-blade, 65
World Governance Indicators (WGIs), 96, 104
WTG. *See* Wind turbine generators (WTG)

Printed in the United States
By Bookmasters